VOLUME EIGHTY-ONE

ADVANCES IN CARBOHYDRATE CHEMISTRY AND BIOCHEMISTRY

Special Volume in Memory of Hidetoshi Yamada Part 1

BOARD OF ADVISORS

VOLUME EIGHTY-ONE

ADVANCES IN CARBOHYDRATE CHEMISTRY AND BIOCHEMISTRY

Special Volume in Memory of Hidetoshi Yamada Part 1

Editor

DAVID C. BAKER
University of Tennessee,
Knoxville, TN

ELSEVIER

ACADEMIC PRESS
An imprint of Elsevier

Academic Press is an imprint of Elsevier
50 Hampshire Street, 5th Floor, Cambridge, MA 02139, United States
525 B Street, Suite 1650, San Diego, CA 92101, United States
The Boulevard, Langford Lane, Kidlington, Oxford OX5 1GB, United Kingdom
125 London Wall, London, EC2Y 5AS, United Kingdom

First edition 2022

ISBN: 978-0-323-98588-8
ISSN: 0065-2318

For information on all Academic Press publications
visit our website at https://www.elsevier.com/books-and-journals

Publisher: Zoe Kruze
Acquisitions Editor: Jason Mitchell
Developmental Editor: Tara Nadera
Production Project Manager:
Vijayaraj Purushothaman
Cover Designer: Mark Rogers

Typeset by STRAIVE, India

Contents

Contributors

Hiromune Ando
Institute for Glyco-Core Research (iGCORE), Gifu University, Gifu, Japan

Kohki Fujikawa
Bioorganic Research Institute, Suntory Foundation for Life Sciences, Kyoto, Japan

Akihiro Imamura
Institute for Glyco-Core Research (iGCORE), Gifu University; Department of Applied Bioorganic Chemistry, Faculty of Applied Biological Sciences, Gifu, Japan

Hideharu Ishida
Institute for Glyco-Core Research (iGCORE), Gifu University; Department of Applied Bioorganic Chemistry, Faculty of Applied Biological Sciences, Gifu, Japan

Yukishige Ito
Graduate School of Science, Osaka University, Toyonaka, Japan

Yasuhiro Kajihara
Department of Chemistry Osaka University, Machikaneyama, Toyonaka, Osaka, Japan

Naoko Komura
Institute for Glyco-Core Research (iGCORE), Gifu University, Gifu, Japan

Yanbo Liu
Department of Chemistry Osaka University, Machikaneyama, Toyonaka, Osaka, Japan

Shoko Mori
Bioorganic Research Institute, Suntory Foundation for Life Sciences, Kyoto; Department of Chemistry, Graduate School of Science, Osaka University, Toyonaka, Japan

Ken-ichi Nishiyama
Department of Biological Chemistry and Food Sciences, Faculty of Agriculture, Iwate University, Morioka, Japan

Kota Nomura
Department of Chemistry Osaka University, Machikaneyama, Toyonaka, Osaka, Japan

Keiko Shimamoto
Bioorganic Research Institute, Suntory Foundation for Life Sciences, Kyoto; Department of Chemistry, Graduate School of Science, Osaka University, Toyonaka, Japan

Hide-Nori Tanaka
Institute for Glyco-Core Research (iGCORE), Gifu University, Gifu, Japan

Shinnosuke Wakamori
Faculty of Life Sciences, Tokyo University of Agriculture, Tokyo, Japan

Preface

Volumes 81 and 82 of *Advances in Carbohydrate Chemistry and Biochemistry* are dedicated to the memory of Professor Hidetoshi Yamada, a chemist, teacher, mentor, and invaluable colleague to a large number of students, research workers, and faculty alike, who has left in his wake a rich legacy of inspiration and guidance. His accomplishments are highlighted in Vol. 81, Chapter 1, "Hidetoshi Yamada: His journey in the carbohydrate world," and Chapter 2, "Yamada's carbohydrate chemistry," respectively, by Yukishige Ito and Shinnosuke Wakamori. Together, the two authors provide insights into the important, creative chemical developments emanating from Prof. Yamada's laboratory. His research targets were highly diverse, including sweet saponins, ellagitannins, novel cyclodextrins, and conformationally distorted donors for stereoselective glycosylations. Chapter 2 concentrates on the development of highly selective glycosylation reactions by bulky silyl-protected or *o*-xylylene-bridged glycosyl donors. The 1,1′-(ethane-1,2-diyl)dibenzene-2,2′-bis(methylene) bridging is known as the "supple conformation system," which, when combined with an α-selective glycosylation protocol, led to the remarkable synthesis of the smallest cyclodextrins on record. From these two chapters, the reader is left impressed with the truly remarkable achievements of Prof. Yamada's research career, though cut short by his untimely passing.

Chapters 3–5 of Vol. 81 (and Chapters 1–5 of Vol. 82) are contributed works by Prof. Yamada's associates and further serve to illustrate the impacts of his works.

In Chapter 3 entitled, "Chemical synthesis of sialoglyco-architectures," Hiromune Ando, Naoko Komura, Hide-Nori Tanaka, Akihiro Imamura, and Hideharu Ishida report on the synthetic developments, particularly for α-sialylated compounds, with a special emphasis on gangliosides and their analogues, including fluorescent probes. Most notable is the development of macrobicyclic sialyl donors that afford α-selective sialylation. These features offer near-complete stereoselectivity, high glycosidation yields, a broad substrate scope, and high applicability to sialic acid congeners that will serve to revolutionize the synthesis of sialoglycans.

Kota Nomura, Yanbo Liu, and Yasuhiro Kajihara in Chapter 4 entitled, "Synthesis of homogeneous glycoproteins with diverse N-glycans," address the post- and cotranslational modifications of proteins that are essential elements for protein function activation. These modifications regulate a wide range of biological processes, including several serious disease states. By combining *Escherichia coli* expression and chemical synthesis, a semisynthesis protocol was developed that provides a methodology for obtaining homogeneous glycoproteins via the facile preparation of a non-glycosylated, long peptide by *E. coli* expression. The full-length glycoprotein can thus be assembled in only a few chemical conversion steps. Robust synthesis of large-sized glycoproteins is a future goal of this highly promising methodology.

In Chapter 5, "A bacterial glycolipid essential for membrane protein integration," Kohki Fujikawa, Shoko Mori, Ken-ichi Nishiyama, and Keiko Shimamoto discuss a new factor, MPIase, involved in the conformation and orientation of membrane protein integration in *E. coli*. Despite its enzyme-like activity, MPIase is a complex glycolipid. Synthesis of a trisaccharyl pyrophospholipid termed "mini-MPIase-3," a minimal unit of MPIase, and its derivatives allowed for structure-activity studies that revealed that MPIase is crucial for membrane protein integration and cell growth. This points to interesting, new biological functions of glycolipids that are probed using a combination of synthetic chemistry, biochemistry, physicochemical measurements, and molecular-biological approaches.

Volume 82 continues with chapters contributed in memory of Prof. Hidetoshi Yamada.

David C. Baker
Knoxville, TN, United States
October 2022

A biographical memoire

Hidetoshi Yamada

CHAPTER ONE

Hidetoshi Yamada: His journey in the carbohydrate world

Yukishige Ito*
Graduate School of Science, Osaka University, Toyonaka, Japan
*Corresponding author: e-mail address: yukito@chem.sci.osaka-u.ac.jp

Contents

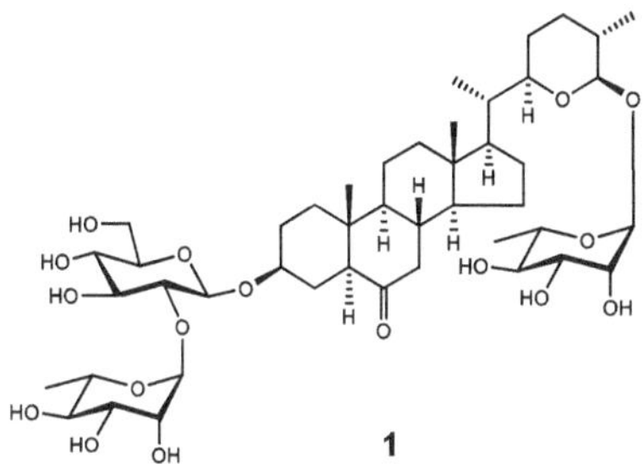

1

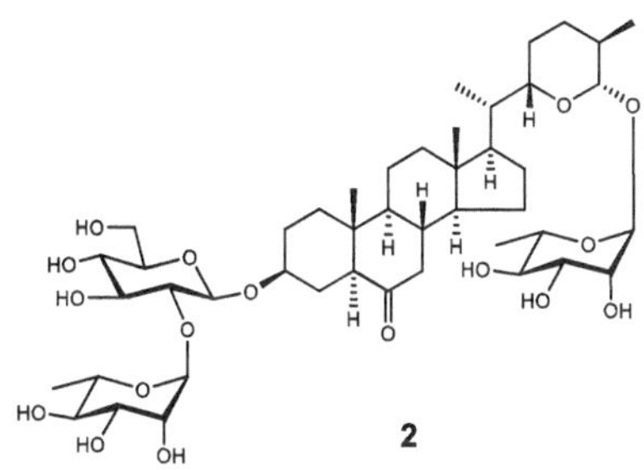

2

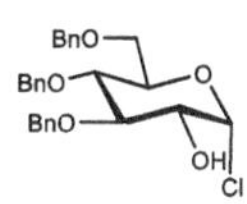

3

Professor Hidetoshi Yamada was born in 1962 and graduated from the Department of Chemistry, Osaka City University in 1985. After obtaining the M.Sci. degree from the same school, he moved to Tokushima Bunri University and continued his research with Professor Mugio Nishizawa, where he received the Ph.D. degree in 1992. After spending a 1-year postdoctoral stay at University of Pittsburgh with Professor Dennis P. Curran, he returned to Tokushima and was promoted to Lecturer in 1996. In 1997, he moved to Kwansei Kaguin University as an Associate Professor and was promoted to Full Professor in 2004. To our deepest sadness, Professor Yamada passed away on 23 November 2019 during a trip to Shikoku Island. This chapter brings to light the most prominent achievements Professor Yamada has made in carbohydrate chemistry.

Advances in Carbohydrate Chemistry and Biochemistry, Volume 81
ISSN 0065-2318
https://doi.org/10.1016/bs.accb.2022.09.001

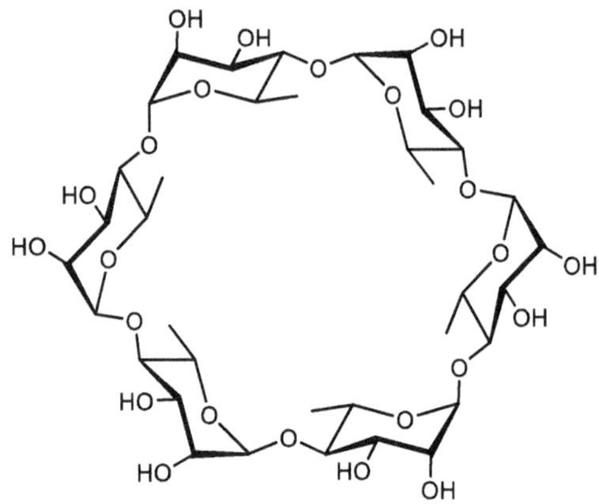

4

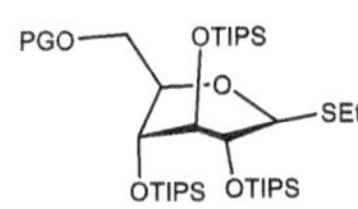

5

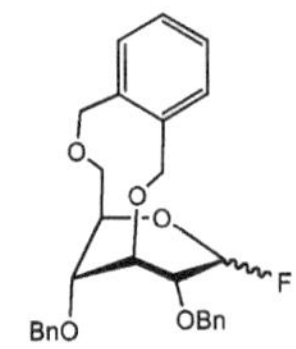

6

Professor Yamada was trained as a synthetic chemist who majored in natural products chemistry and made landmark achievements in carbohydrate chemistry in various respects.[1] His interest in carbohydrate chemistry was initiated when he selected sweet saponins as synthetic targets.[2–5] In particular, through the total synthesis of osladin, he proved that the previously proposed structure **2** was a diastereomer of the natural substance **1**.[4,5] Quite astonishingly, the diastereomer **2**, the previously believed structure of osladin, did not exhibit any taste, while the real osladin was intensively sweet. Interestingly, he employed 2-*O*-unprotected glycosyl chloride **3** to introduce β-linked glucose, a strategy that had been rarely used for the construction of 1,2-trans glycosides. His unique perspective on carbohydrates was reflected in the development of "thermal" glycosylation that obviates the need for typical promoters such as a metal salt or Lewis acid, yet gives the desired glycosides with a high degree of α-selectivity in the presence of an acid scavenger such as α-methylstyrene or tetramethylurea.[6] This chemistry was applied to the synthesis of novel cyclic oligosaccharides **4** composed of L-rhamnopyranose,[7] which was streamlined by using a thioglycoside primer that was elongated by successive thermal glycosylations.[8]

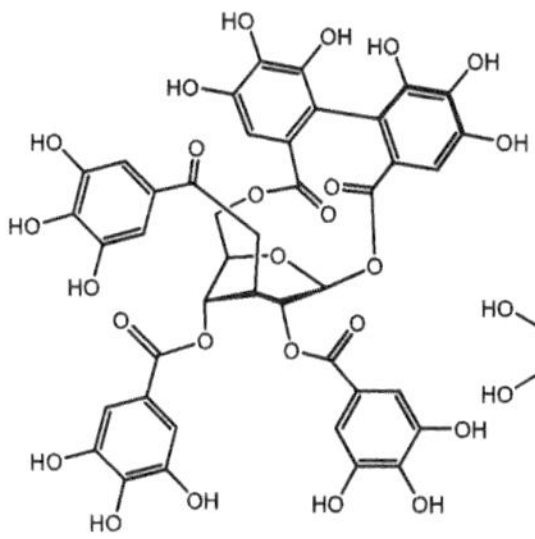

7

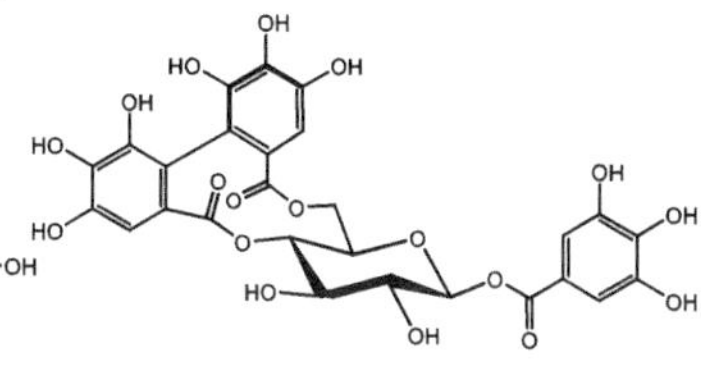

8

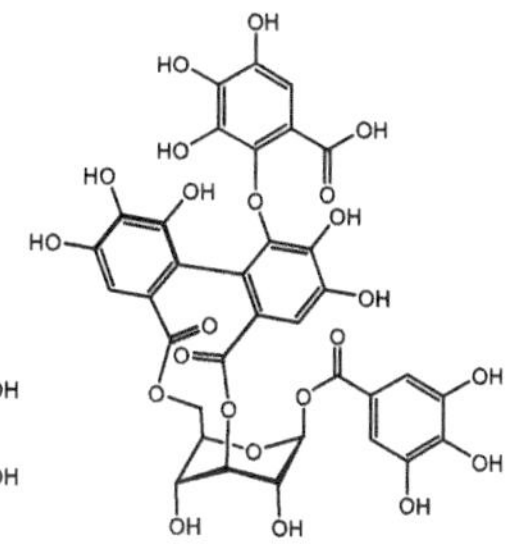

9

Among various achievements of Professor Yamada, the most prominent would be the development of glycosyl donors that adopt uncommon conformation. He first discovered conformational distortion of hexopyranoses to a ring-inverted chair (${}^{1}C_4$) or a twist-boat conformation when multiply protected by sterically demanding silyl groups.[9,10] As an example, 2,3,4-tri-O-TIPS protected thioglucoside **5** adopts the twist-boat conformation, which exhibits high β-selectivity when used as a glycosyl donor.[10–12]

He explored an even more sophisticated approach toward achieving the conformational control of glycosylation by bridging two positions of glycosyl donors. Specifically, he introduced the bridge between the 3- and 6-positions by an *o*-xylylene group, forcing the glycosyl donor to adopt the axial-rich conformation.[13] The donor **6** exhibited complete β-selectivity toward various acceptors, while the conditions that enable α-selective glycosylation were also discovered.[14]

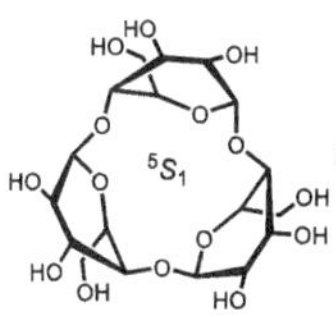

10

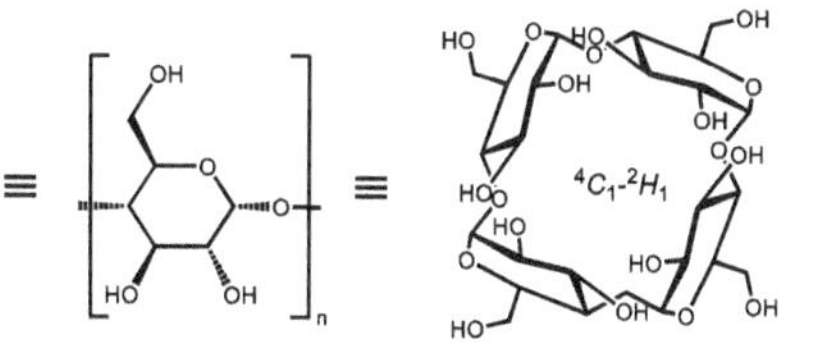

11

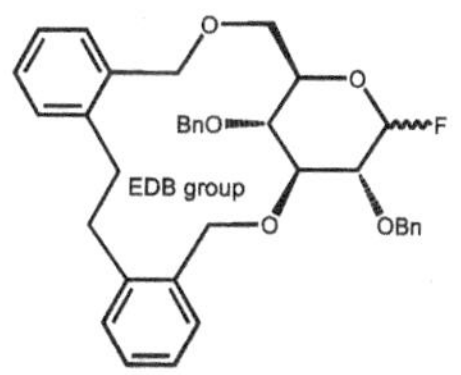

12

Professor Yamada fascinated his followers by a series of studies toward the total synthesis of ellagitannins,[15–28] a group of natural products having a carbohydrate component bridged by a CC-dimer of galloyl groups such as (+)-davidiin (**7**)[17] and strictinin (**8**).[18] A most recent example, the total synthesis of macaranin B (**9**), which is bridged by a macaranoyl group including tetra-ortho-substituted diaryl structures, was achieved.[28]

The concept of bridging sugar skeleton culminated in the synthesis of the smallest cyclodextrins **10** and **11** consisting of 3 and 4 glucose units from the

3,6-*O*-EDB bridged donor **12**, a landmark achievement in the history of carbohydrate chemistry.[29]

Professor Yamada's achievements demonstrated his unparalleled originality. We can see clearly that, throughout his whole career, he had been solely driven by curiosity and free from short-sighted trends. I believe his spirit will be passed on to the next generations in both the natural products and carbohydrate communities.

Acknowledgment

I wish to thank Dr. Shinnosuke Wakamori for his valuable comments.

References

1. https://orcid.org/0000-0003-2272-319X.
2. Yamada, H.; Nishizawa, M. Synthesis of Sweet Tasting Diterpene Glycosides, Baiyunoside and Analogs. *Tetrahedron* **1992**, *48*, 3021–3044. https://doi.org/10.1016/S0040-4020(01)92246-X.
3. Yamada, H.; Nishizawa, M. Total Synthesis of Baiyunoside by a Novel 2' Discriminated Glycosidation. *Tetrahedron Lett.* **1987**, *28*, 4315–4318. https://doi.org/10.1016/S0040-4039(00)96495-5.
4. Yamada, H.; Nishizawa, M. Total Synthesis of Intensely Sweet Saponin, Osladin. *Synlett* **1993**, 54–56. https://doi.org/10.1055/s-1993-22346.
5. Yamada, H.; Nishizawa, M. Total Synthesis and Structure Revision of Intensely Sweet Saponin Osladin. *J. Organomet. Chem.* **1995**, *60*, 386–397. https://doi.org/10.1021/jo00107a018.
6. Nishizawa, M.; Kan, Y.; Yamada, H. Thermal Glycosidation with Benzylated Glycosyl Chlorides: A Very Simple Procedure of Glycosidation. *Chem. Pharm. Bull.* **1989**, *37*, 565–567.
7. Nishizawa, M.; Imagawa, H.; Kan, Y.; Yamada, H. Total Synthesis of *cyclo*-L-Rhamnohexaose by a Stereoselective Thermal Glycosylation. *Tetrahedron Lett.* **1991**, *32*, 5551–5554. https://doi.org/10.1016/0040-4039(91)80081-G.
8. Nishizawa, M.; Imagawa, H.; Kubo, K.; Kan, Y.; Yamada, H. Improved Synthesis of α-Cycloawaodorin. *Synlett* **1992**, 447–448. https://doi.org/10.1055/s-1992-21376.
9. Yamada, H.; Tanigakiuchi, K.; Nagao, K.; Okajima, K.; Mukae, T. The First Ring Inversion of Pyranoses Induced by Bulky Silyl Protections at the 2- and 3-Positions. *Tetrahedron Lett.* **2004**, *45*, 9207–9209. https://doi.org/10.1016/j.tetlet.2004.10.078.
10. Yamada, H.; Nakatani, M.; Ikeda, T.; Marumoto, Y. Stable Axial-Rich Conformation of Pyranoses Derived from L-Rhamnose and D-Mannose. *Tetrahedron Lett.* **1999**, *40*, 5573–5576. https://doi.org/10.1016/S0040-4039(99)01077-1.
11. Okada, Y.; Nagata, O.; Taira, M.; Yamada, H. Highly β-Selective and Direct Formation of 2-O-Glycosylated Glucosides by Ring Restriction into Twist-Boat. *Org. Lett.* **2007**, *9*, 2755–2758. https://doi.org/10.1021/ol070720b.
12. Okada, Y.; Mukae, T.; Okajima, K.; Taira, M.; Fujita, M.; Yamada, H. Highly β-selective O-Glucosidation Due to the Restricted Twist Boat Conformation. *Org. Lett.* **2007**, *9*, 1573–1576. https://doi.org/10.1021/ol070427b. Mar 2007.
13. Okada, Y.; Asakura, N.; Bando, M.; Ashikaga, Y.; Yamada, H. Completely β-Selective Glycosylation using 3,6-*O*-(*o*-Xylylene)-Bridged Axial-Rich Glucosyl Fluoride. *J. Am. Chem. Soc.* **2012**, *134*, 6940–6943. https://doi.org/10.1021/ja301480g.

14. Motoyama, A.; Arai, T.; Ikeuchi, K.; Aki, K.; Wakamori, S.; Yamada, H. α-Selective Glycosylation of 3,6-*O*-*o*-Xylylene-Bridged Glucosyl Fluoride. *Synthesis* **2018**, *50*, 282–294. https://doi.org/10.1055/s-0036-1590927.
15. Ikeda, Y.; Nagao, K.; Tanigakiuchi, K.; Tokumaru, G.; Tsuchiya, H.; Yamada, H. The First Construction of a 3,6-Bridged Ellagitannin Skeleton with $^1C_4/B$ Glucose Core, Synthesis of Nonamethylcorilagin. *Tetrahedron Lett.* **2004**, *45*, 487–489. https://doi.org/10.1016/j.tetlet.2003.11.006. 2-319X.
16. Yamada, H.; Nagao, K.; Dokei, K.; Kasai, Y.; Michihata, N. Total Synthesis of (−)-Corilagin. *J. Am. Chem. Soc.* **2008**, *130*, 7566–7567. https://doi.org/10.1021/ja803111z.
17. Kasai, Y.; Michihata, N.; Nishimura, H.; Hirokane, T.; Yamada, H. Total Synthesis of (+)-Davidiin. *Angew. Chem. Int. Ed.* **2012**, *51*, 8026–8029. https://doi.org/10.1002/anie.201203305.
18. Michihata, N.; Kaneko, Y.; Kasai, Y.; Tanigawa, K.; Hirokane, T.; Higasa, S.; Yamada, H. High-Yield Total Synthesis of (−)-Strictinin Through Intramolecular Coupling of Gallates. *J. Organomet. Chem.* **2013**, *78*, 4319–4328. https://doi.org/10.1021/jo4003135.
19. Yamaguchi, S.; Hirokane, T.; Yoshida, T.; Tanaka, T.; Hatano, T.; Ito, H.; Nonaka, G.; Yamada, H. Roxbin B Is Cuspinin; Structural Revision and Total Synthesis. *J. Organomet. Chem.* **2013**, *78*, 5410–5417. https://doi.org/10.1021/jo400562k.
20. Ashibe, S.; Ikeuchi, K.; Kume, Y.; Wakamori, S.; Ueno, Y.; Iwashita, T.; Yamada, H. Non-Enzymatic Oxidation of a Pentagalloylglucose Analog to Ellagitannins. *Angew. Chem. Int. Ed.* **2017**, *56*, 15402–15406. https://doi.org/10.1002/anie.201708703.
21. Yamashita, K.; Kume, Y.; Ashibe, S.; Puspita, C. A. D.; Tanigawa, K.; Michihata, N.; Wakamori, S.; Ikeuchi, K.; Yamada, H. Total Synthesis of Mallotusinin. *Chem. Eur. J.* **2020**, *26*, 16408–16421. https://doi.org/10.1002/chem.202002753.
22. Wakamori, S.; Matsumoto, S.; Kusuki, R.; Ikeuchi, K.; Yamada, H. Total Synthesis of Casuarinin. *Org. Lett.* **2020**, *22*, 3392–3396. https://doi.org/10.1021/acs.orglett.0c00876.
23. Yamada, H.; Ohara, K.; Ogura, T. Total Synthesis of Cercidinin A. *Eur. J. Org. Chem.* **2013**, 7872–7875. https://doi.org/10.1002/ejoc.201301219.
24. Hirokane, T.; Hirata, Y.; Ishimoto, T.; Nishii, K.; Yamada, H. A Unified Strategy for the Synthesis of Highly Oxygenated Diaryl Ethers Featured in Ellagitannins. *Nat. Commun.* **2014**, *5*. https://doi.org/10.1038/ncomms4478.
25. Hirokane, T.; Ikeuchi, K.; Yamada, H. Total Syntheses of Laevigatins A and E. *Eur. J. Org. Chem.* **2015**, 7352–7359. https://doi.org/10.1002/ejoc.201501037.
26. Kaneko, Y.; Wakamori, S.; Ikeuchi, K.; Ohara, K.; Tanaka, T.; Yamada, H. Total Synthesis of Lagerstannin C: Follow-Up of the Khanbabaee's Synthesis. *Synthesis* **2017**, *49*, 5003–5006. https://doi.org/10.1055/s-0036-1588477.
27. Ikeuchi, K.; Ueji, T.; Matsumoto, S.; Wakamori, S.; Yamada, H. First Total Synthesis of Neostrictinin. *Eur. J. Org. Chem.* **2020**, 2077–2085. https://doi.org/10.1002/ejoc.202000053.
28. Matsumoto, S.; Aoyama, A.; Wakamori, S.; Yamada, H. Total Synthesis of Macaranin B. *Biosci. Biotechnol. Biochem.* **2021**, *85*, 1937–1944. https://doi.org/10.1093/bbb/zbab120.
29. Ikuta, D.; Hirata, Y.; Wakamori, S.; Shimada, H.; Tomabechi, Y.; Kawasaki, Y.; Ikeuchi, K.; Hagimori, T.; Matsumoto, S.; Yamada, H. Conformationally Supple Glucose Monomers Enable Synthesis of the Smallest Cyclodextrins. *Science* **2019**, *364*, 674–677. https://doi.org/10.1126/science.aaw3053.

CHAPTER TWO

Yamada's carbohydrate chemistry

Shinnosuke Wakamori*
Faculty of Life Sciences, Tokyo University of Agriculture, Tokyo, Japan
*Corresponding author: e-mail address: sw207675@nodai.ac.jp

Contents

1. Introduction

Professor Hidetoshi Yamada studied chemistry at Osaka City University where he obtained a Bachelor's degree in Chemistry (1985). After receiving a Master's degree in Chemistry (1987) from the same university, he became Assistant Professor at Tokushima Bunri University where he worked in the laboratory of Professor Mugio Nishizawa. After gaining the position of Associate Professor at the School of Science and Technology of Kwansei Gakuin University in 1997, he attained the position of Full Professor in 2004.

The concept of the steric conformation in sugar chemistry was a driving force in Professor Yamada's scientific research. The trigger for the research was the synthetic studies of osladin (**1**) and α-cycloawaodorin (**2**) that he conducted at Tokushima Bunri University (Fig. 1).[1,2] Because these syntheses required the α-selective glycosylation of L-rhamnose, he began to develop an efficient glycosylation reaction. In biological systems, enzymes deform the steric conformation of D-glucose, leading to selective glycosylation reactions. Thus, his study on the development of a stereoselective glycosylation reaction was focused on deforming the glucopyranose ring from the most stable chair conformation. In 2000, an axial-rich chair system utilizing L-rhamnose derivative **3** was reported. Seven years later, his studies

Advances in Carbohydrate Chemistry and Biochemistry, Volume 81
ISSN 0065-2318
https://doi.org/10.1016/bs.accb.2022.09.002

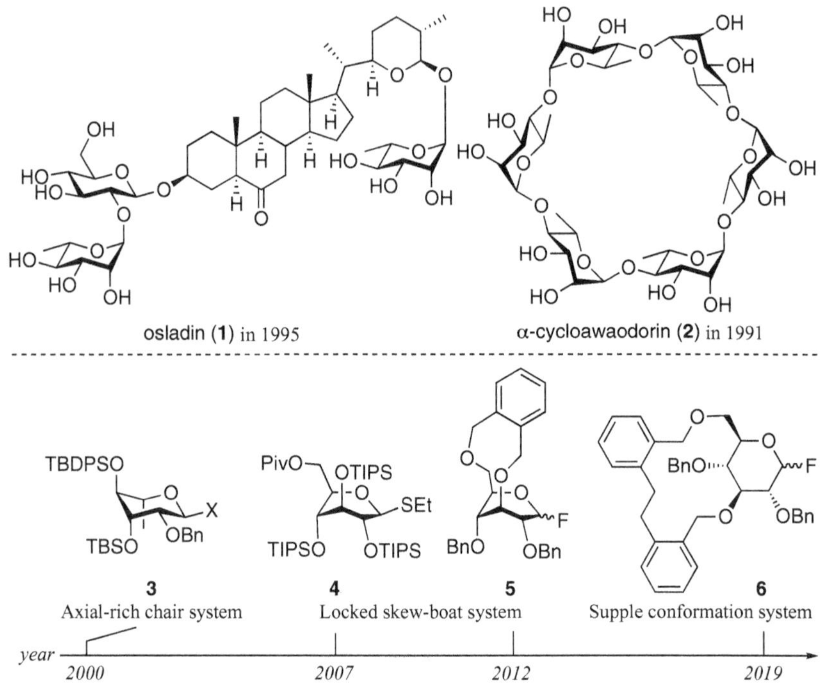

Fig. 1 Overview of Professor Yamada's research timeline.

expanded the axial-rich chair system to a locked skew-boat system, and selective glycosylation reactions were developed with glycosyl donors **4** and **5**. In 2019, his research led to a breakthrough discovery of a supple conformation system of the glucopyranose ring that led to the synthesis of the smallest cyclodextrins through the specially designed glycosyl donor **6**. In this chapter, the research trajectory conducted by Professor Yamada is described in chronological order.

2. Axial-rich chair system

In 2000, Professor Yamada in his research began to restrict the steric conformation of L-rhamnose (Fig. 2).[3,4] Introduction of a *tert*-butyldimethylsilyl (TBS) group into O-3 and a *tert*-butyldiphenylsilyl (TBDPS) group into the O-4 of L-rhamnose (**7**) restricted the ring-conformation of **3** into an axial-rich chair form, a $^4C_1$ conformation. The characteristic α-preference then disappeared in the rhamnosylation reaction when utilizing phenyl 1-thio-*α*-L-rhamnopyranoside or L-rhamnopyranosyl fluoride (α/β = 47/53 and 57/43, respectively). The application of a bulkier Lewis acid increased

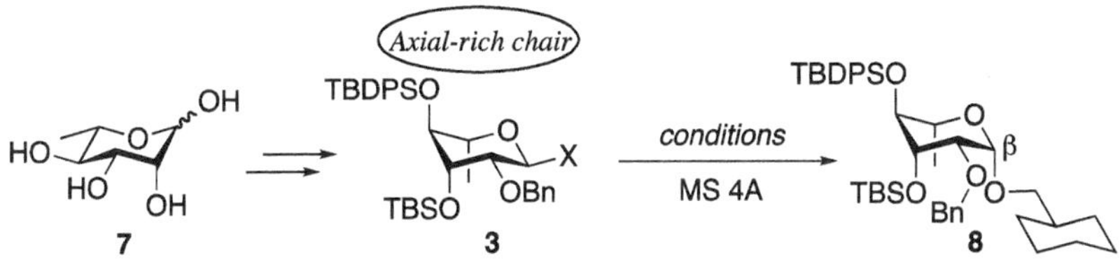

X	activator	solvent	temp.	yield of 8	α/β
–SPh	NBS	THF	0 °C	68%	47/53
–F	$SnCl_2$, $AgClO_4$	Et_2O	−15 °C	84%	57/43
–O–C(=NH)–CCl_3	9-BBNOTf	hexane	−95 °C	73%	27/73

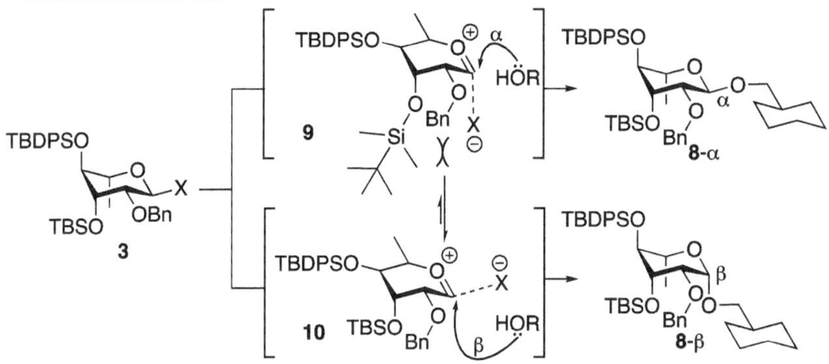

Fig. 2 Glycosylation study of axial-rich chair formed L-rhamnosyl donor and proposed reaction mechanism.

the β-selectivity. Ultimately, using of 9-borabicyclo[3.3.1]nonyl trifluoromethanesulfonate (9-BBNOTf) and lowering the reaction temperature with α-L-rhamnopyranosyl trichloroimidate yielded a 73% of β-selectivity (α/β = 27/73). In the glycosylation, the oxocarbenium ion intermediate should be at equilibrium between **9** and **10**, in which **10** would be more stable. With bulkier Lewis acids, the steric repulsion increased in **9** due to additional repulsion between the Lewis acid and the TBS group at O-3. These results show that the latent α-preference of L-rhamnosylation disappeared and revealed that the conformational change of a sugar greatly affects the stereoselectivity in the glycosylation reaction.

The conformation of the glucopyranose ring while introducing the TBS or TBDPS groups into the two adjacent hydroxy groups on D-glucose was also studied (Fig. 3).[5,6] TBDPS groups were sufficiently bulky to restrict the ring conformation to the axial-rich chair form by introducing bulky silyl

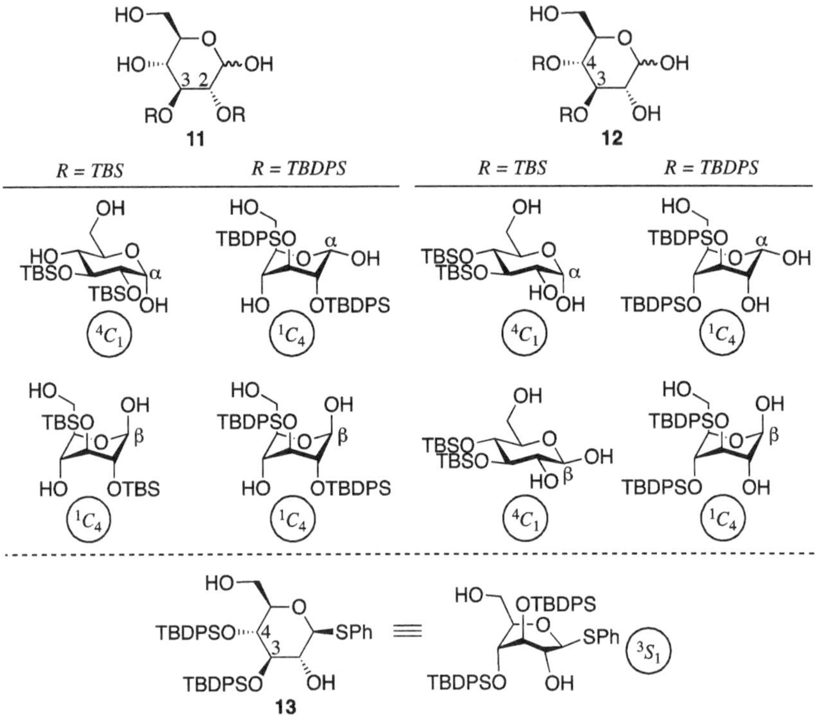

Fig. 3 Conformational study of axial-rich chair formed D-glucose derivative by introduction of various silyl groups.

protecting groups onto adjacent *trans*-diols, but TBS groups were not. Indeed, the introduction of the TBDPS groups restricted the ring into the axial-rich chair conformation whether the protected positions were in O-2 and O-3 (for **11**) or in O-3 and O-4 (for **12**). In contrast, the glucopyranose ring of **13** possessing TBDPS groups at O-3 and O-4 and a 1β-phenylthio group existed in a skew-boat type form, a $^3S_1$ conformation. This discovery led to the development of a method to lock the glucopyranose ring in the skew-boat form.

3. Locked skew-boat system

In 2007, the locked skew-boat system allowed a highly β-selective glycosylation (Fig. 4).[7] As protection of O-2, O-3, and O-4 with triisopropylsilyl (TIPS) groups, the ring conformation of **4** was in a $^3S_1$ conformation. NMR studies revealed that the methylene group at C-6 was oriented in the equatorial direction. In the glycosylation using the

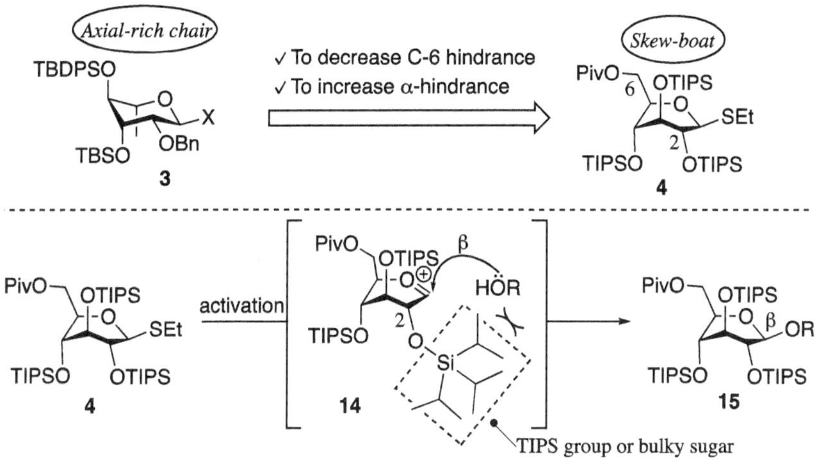

Fig. 4 Development of a locked skew-boat system by protection with three TIPS groups at O-2, O-3, and O-4 on the D-glucopyranose ring.

axial-rich chair derivative, the axially oriented substituent at O-2 blocks the approach of a nucleophile from the α face, whereas the axially oriented the functional group at O-3 and the methylene group at C-6 make double 1,3-diaxial relationships against the approaching nucleophiles. In contrast, for the skew-boat conformation, there was no steric repulsion due to C-6, which was the one of the assumed double 1,3-diaxial repulsions when the alcohol approached from the β face. Therefore, for the steric repulsion of the reaction intermediate **14**, the approach of the alcohol was against the huge TIPS group at O-2 in the α face. High β-selectivity was maintained even when a bulky sugar was attached to O-2 instead of the TIPS group.

Using **4** as the glycosyl donor and methyl trifluoromethanesulfonate (MeOTf) as the activator in dichloromethane achieved a highly β-selective glycosylation (Fig. 5).[7] The protecting group at O-6 was employed as a pivaloyl group, which gave glucosides with quite high β-selectivity. These reaction conditions were also applicable to the case where various secondary and tertiary alcohols were used as glycosyl acceptors, giving the corresponding *O*-β-D-glucosides **15** in a highly β-selective manner. The reaction with cyclohexylmethanol gave the corresponding glucoside in a 92% yield and α/β ratio of 5/95. Reactions with (+)-menthol and (−)-menthol gave glucosides with comparable β-selectivity regardless of their chirality, indicating that the possible matching-mismatching effect[8,9] between the glycosyl donor and acceptor did not affect this reaction system

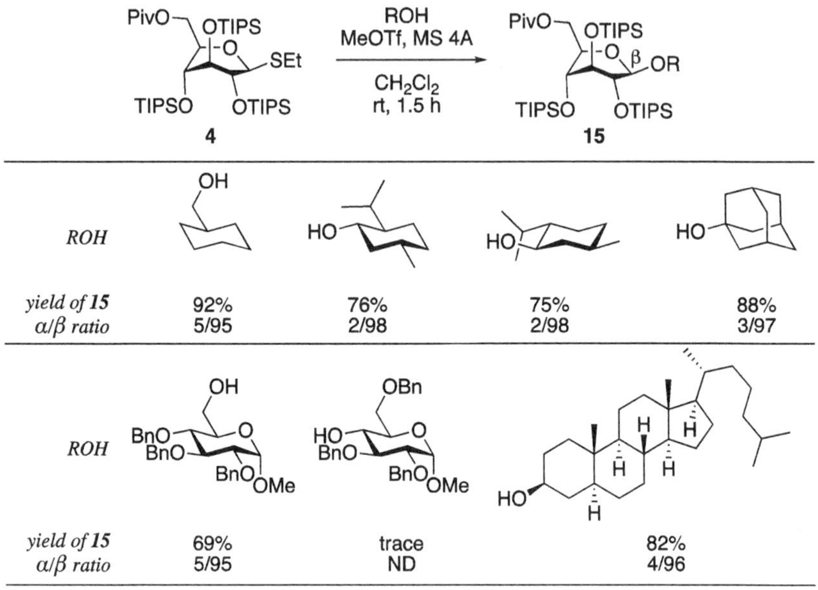

Fig. 5 β-Selective glycosylation of tri-TIPS protected glycosyl donor with various alcohols.

($\alpha/\beta = 2/98$ each). Furthermore, the reactions with 1-adamantanol, a tertiary alcohol, cholestanol, a bulky molecule as a whole, and methyl 2,3,4-tri-*O*-benzyl-α-D-glucopyranoside derivative proceeded similarly. In contrast, the reaction with methyl 2,3,6-tri-*O*-benzyl-α-D-glucopyranoside yielded a glycal generated from **4** as the main product, with only a trace of desired glucoside **15**. Additionally, migration of the silyl protecting group on the glucosyl donor to the hydroxyl group of the glycosyl acceptor was observed. The steric configuration of products **15** was confirmed on the basis of ^{1}H NMR coupling constants after all the products were desilylated to gain the $^4C_1$ conformation of glucopyranose.

The β-selective glycosylation using the locked skew-boat system was particularly effective for the synthesis of 2-*O*-glycosylated *O*-β-D-glucosides (Fig. 6).[10] The *O*-β-D-glucoside with another sugar attached to O-2 is a structure that is often present in simple aliphatic hydrocarbons as well as natural glycosides with coumarins, flavones, terpenes, and steroids as aglycons. 2-*O*-Glycosylated *O*-β-D-glucosides are one of the typical structural patterns that have been difficult to prepare using classical neighboring-group participation. The method described here is a convergent synthetic

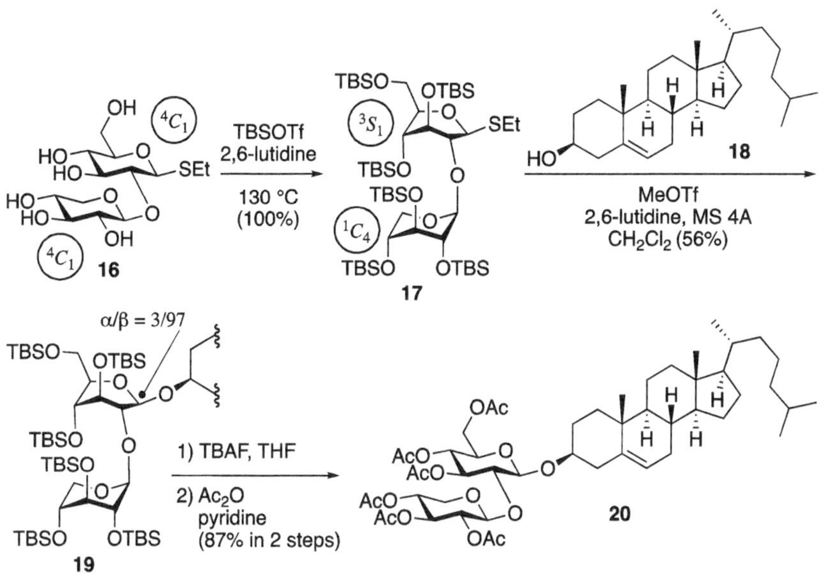

Fig. 6 Application of skew-boat system to 2-*O*-glycosylated *O*-β-D-glucosides.

route that can be adopted when the 2-*O*-glycosylated-glucosyl donor is introduced directly with high β-selectivity. Heating the mixture of 2-*O*-xylosylated thioglycoside **16**, TBSOTf, and 2,6-lutidine at 130 °C protected all hydroxy groups with the TBS groups to restrict the glucopyranose and xylopyranose rings of **17** into $^3S_1$ and $^1C_4$ conformations, respectively. The β-selective glycosylation of disaccharide donor **17** with cholesterol (**18**) yielded desired glycoside **19** in a ratio of 3/97. Treatment of **19** with tetrabutylammonium fluoride (TBAF) removed all the TBS groups. Subsequent acetylation of the six hydroxy groups then afforded the corresponding hexaacetate **20** in an 87% yield. Even when the second sugar moiety (D-xylose) of **16** was replaced with D-glucose or L-rhamnose, the ring-conformations of the glucose parts of **17** remained similarly in the skew-boat conformation. These analogues also afforded the β-selective direct introduction of the disaccharides.

The skew-boat system was also applied to natural product synthesis.[11] Davidiin (**21**) is an ellagitannin, a class of polyphenol, isolated from *Davidia involucre* (Fig. 7).[12,13] The compound inhibits the binding of a ligand to a μ-opioid receptor.[14] Structurally, **21** contains galloyl groups at O-2, O-3, and O-4 and an (*S*)-hexahydroxydiphenoyl (HHDP) group between O-1 and O-6 that deforms the glucopyranose ring to the $^3S_1$ conformation.

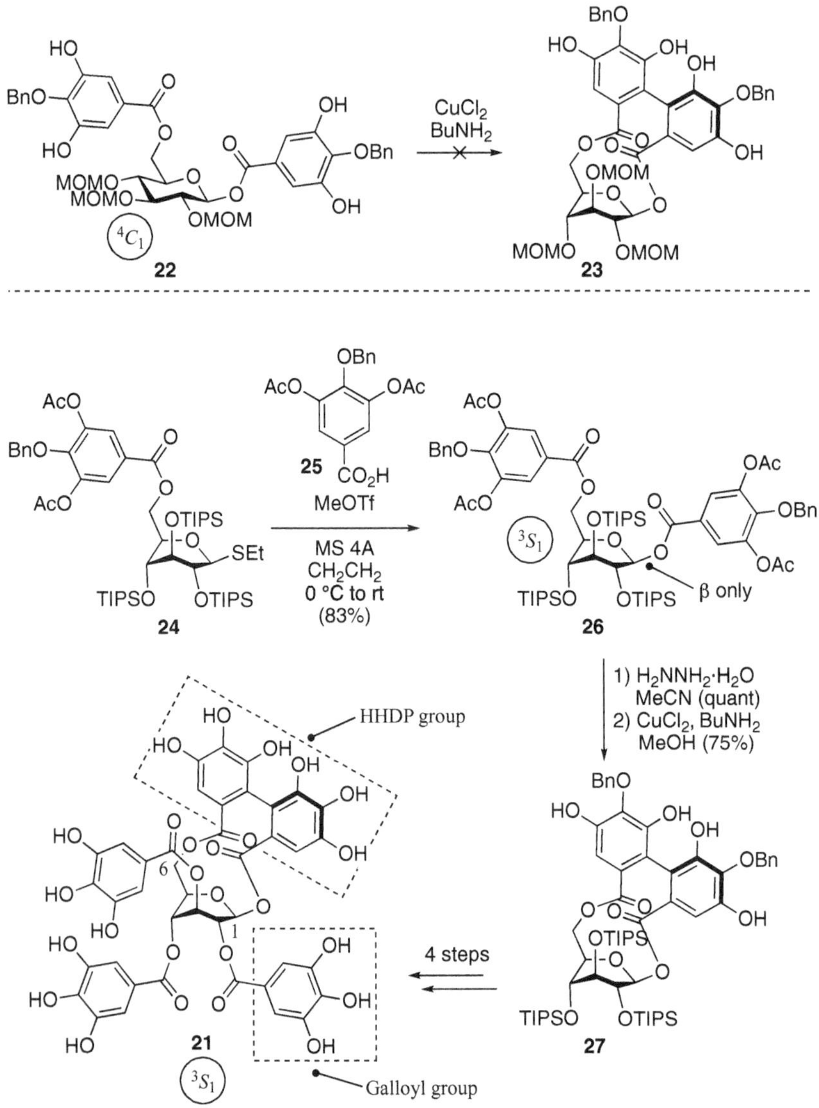

Fig. 7 Application of skew-boat system to natural product synthesis.

The formation of the 1,6-*O*-(*S*)-HHDP bridge was impossible, because it required simultaneous conformational inversion of the pyranose ring into the thermodynamically unfavorable axial-rich chair form. This problem became evident during an attempt to carry out the intramolecular oxidative coupling of galloyl groups on **22** with the stable $^{4}C_{1}$ conformation. Treatment of **22** with the copper(II) chloride/butylamine system, a reagent combination that effectively coupled 4-*O*-benzylated gallates,[15] did not produce the desired product **23**. Therefore, both the β-selective glycosylation reaction and the steric conformation deformation of the skew-boat system were employed in the synthesis of davidiin (**21**). Treatment of **24** with MeOTf in the presence of carboxylic acid **25** as the nucleophile provided **26** in an 83% yield as a single diastereomer. The conformation of the glucopyranose ring of **26** was in the $^{3}S_1$ form, the same as that of davidiin (**21**). After removal of four acetyl groups with hydrazine monohydrate, the oxidative coupling reaction[15] of two galloyl groups successfully constructed the 1,6-*O*-(*S*)-HHDP group to afford **27** as a single diastereomer in 75% yield. The total synthesis of davidiin (**21**) was achieved after a four-step conversion as follows[1]: the protection of four hydroxy groups for benzyl groups,[2] the removal of three TIPS groups,[3] the esterification of three galloyl groups, and[4] the hydrogenolysis of all benzyl groups. The spectroscopic data of synthetically obtained **21** were identical to those of the naturally occurring **21**.[13,16]

In 2012, a 3,6-*O*-*o*-xylylene bridged glycosyl donor was developed (Fig. 8).[17] The conformational control of the glucopyranose using the silyl groups was a factor closely related to stereoselectivity in glycosylation reactions. However, as the steric hindrance around the hydroxyl group of the glycosyl acceptor increased, the yield of the glycosylation reaction decreased. Indeed, the method was not applied to the reaction with the secondary hydroxyl group of methyl 2,3,6-tri-*O*-benzyl-α-D-glucopyranoside. Furthermore, silyl groups were easily transferred under both acidic and basic conditions.[18] Therefore, the development of a more stable method of locking the conformation than the use of steric repulsion of the silyl groups became necessary. The deforming of the pyranose ring to the skew-boat conformation by bridging two non-adjacent hydroxyl groups is often found in natural ellagitannins.[19,20] Based on the structure of ellagitannins, a glycosyl fluoride **5** was designed as a glycosyl donor by bridging the hydroxyl groups between O-3 and O-6 of the glucopyranose ring with an easily removable *o*-xylylene group.[21] The glycosylation with the 3,6-*O*-*o*-xylylene bridged D-glucosyl fluoride **5** achieved high α- or β-selectivities. The activation of **5** generated the corresponding oxocarbenium ion intermediate **28**, in which the *o*-xylylene bridge stood upright to the β-face of the glucopyranose ring and became a steric hindrance for the approach of the alcohol. Therefore, the alcohol attacked from the α-face to afford **29**-α. The thermodynamically controlled reaction consisted of glycosylation and isomerization cycles, catalyzed respectively by $SnB(C_6F_5)_4Cl$[22]

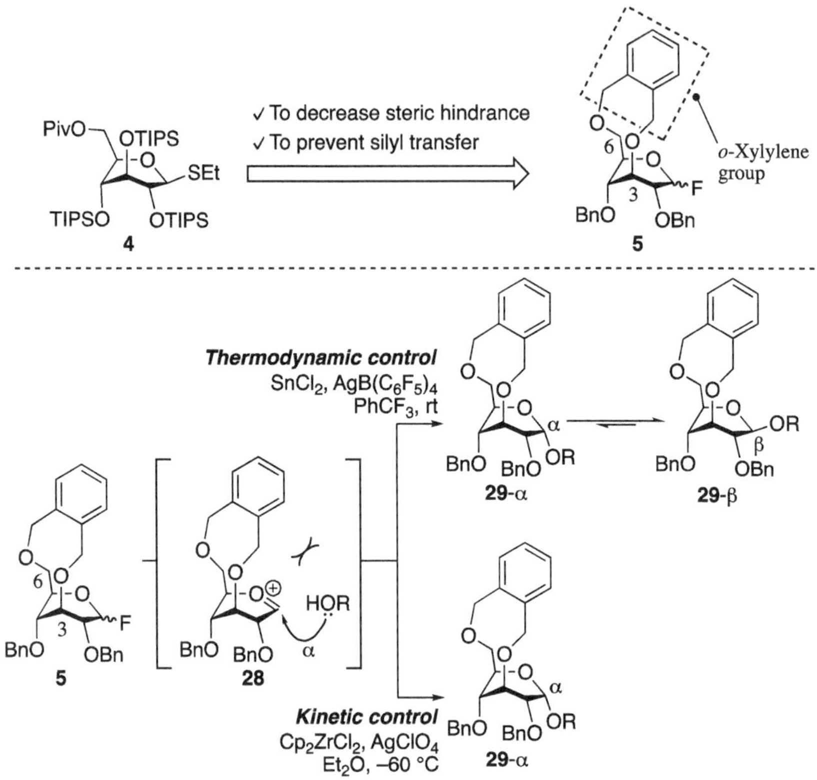

Fig. 8 Development of locked skew-boat system by protection with an o-xylylene bridging group between O-3 and O-6 on the D-glucopyranose ring.

and $HB(C_6F_5)_4$, both generated in situ. The 1,2-*trans*-selectivity arose thermodynamically in the isomerization cycle of **29**-α and **29**-β because of the steric repulsion between O-1α and O-2. Hence, **29**-β was obtained in a highly selective manner. In contrast, to obtain high α-selectivity exploiting the steric hindrance over the β face, kinetic reaction conditions were required because the corresponding glucosides converge to the β-isomer under thermodynamic conditions. Thus, the kinetic conditions with the use of Cp_2ZrCl_2 (where Cp is cyclopentadienyl) and $AgClO_4$ (Suzuki glycosylation)[23] was developed, yielding **29**-α with high α-selectivity.

The thermodynamic glycosylation of 3,6-*O*-*o*-xylylene-bridged glycosyl fluoride enabled a perfect β-stereoselective glycosylation (Fig. 9).[17] The perfect β-stereoselectivity observed in all cases was noteworthy; the α-isomers of **29**-β were not detected by ^{1}H NMR analysis of the crude products. The reaction with cyclohexylmethanol gave the corresponding

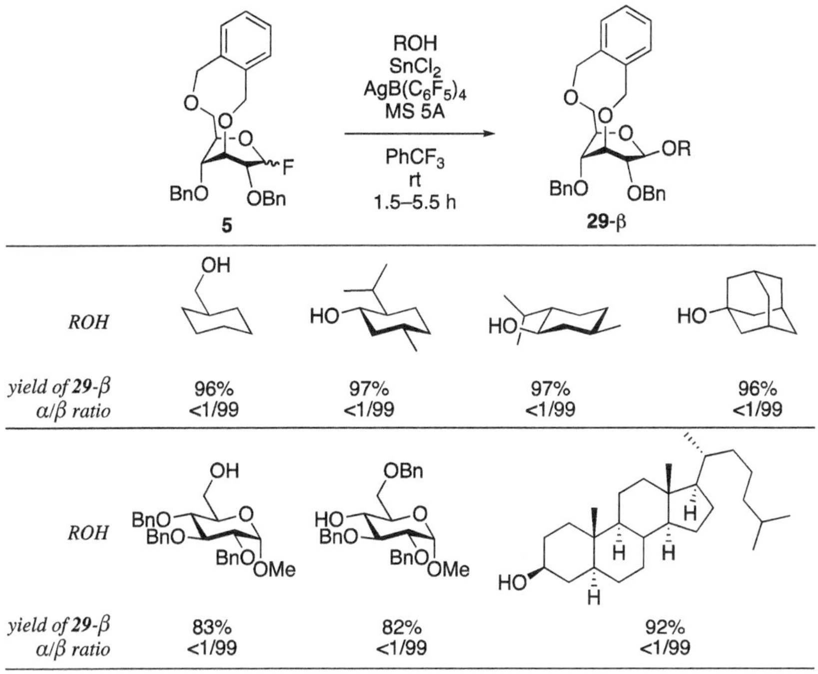

Fig. 9 Thermodynamic β-selective glycosylation of *o*-xylylene-bridged glycosyl donor with various alcohols.

glycoside in a 96% yield. The reactions with (+)-menthol and (–)-menthol showed similar 97% yields and β-stereoselectivities, indicating that there was no matching–mismatching effect[8,9] as well as tri-TIPS protected glycoside **4**. In most cases, the yields were higher than those of the method using silyl group-based conformational control described in the previous section. Furthermore, 3,6-*O*-*o*-xylylene-bridged glycosyl fluoride **5** was applied to a reaction with a very crowded the hydroxyl group at O-4, due to the O-benzyl groups on methyl 2,3,6-tri-*O*-benzyl-α-D-glucopyranoside, to provide **29**-β in an 82% yield.

Under kinetic control conditions, glycosylation of 3,6-*O*-*o*-xylylene bridged glycosyl fluoride **5** showed a high α-selectivity (Fig. 10).[24] Using **5** as a glycosyl donor under Suzuki's reaction conditions[23] generated the α-selective glycosylation. The reaction with cyclohexylmethanol provided the corresponding glucoside **29**-α in a 93% yield and α/β ratio of 97/3. In the reactions with each enantiomer of menthol, a difference in stereo-selectivity was observed. The reaction with (+)-menthol proceeded with higher α-selectivity (α/β=89/11) than the reaction with the (–)-isomer

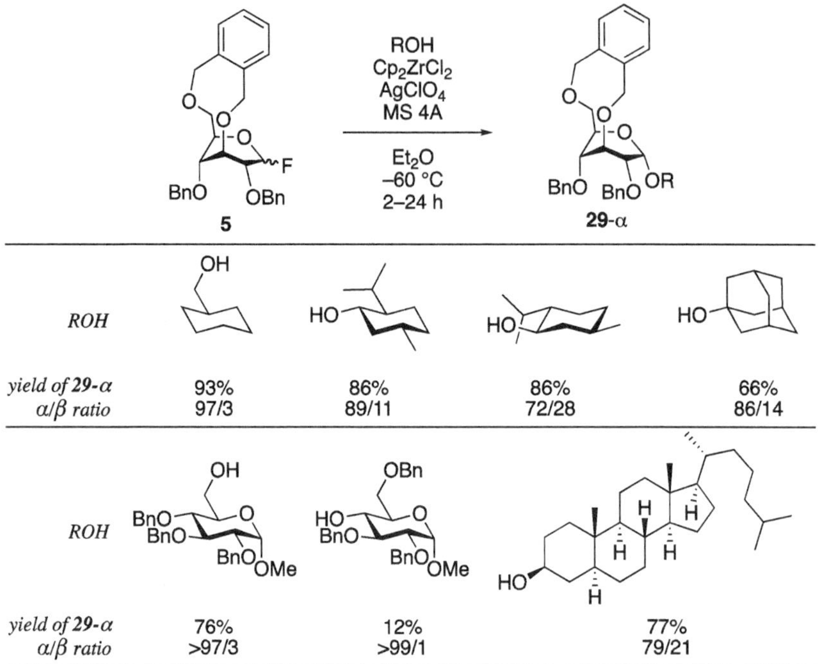

Fig. 10 Kinetic α-selective glycosylation of *o*-xylylene bridged glycosyl donor with various alcohols.

(α/β = 72/28). The matching and mismatching effect of the enantiomers could explain the difference.[8,9] The reaction with a tertiary alcohol, 1-adamantanol, proceeded in a moderately α-selective manner (α/β = 86/14). The reaction at 6-OH of the D-glucose derivative, methyl 2,3,4-tri-*O*-benzyl-α-D-glucopyranoside, provided the corresponding disaccharide in a 76% yield with a >97:3 α/β ratio. In contrast, the hydroxy group at O-4 of a D-glucose derivative, 2,3,6-tri-*O*-benzyl-α-D-glucopyranoside, reacted with a low isolated yield (12%), while the α-selectivity was perfect. For the reaction with cholestanol, the α/β ratio was moderate (79/21).

4. Supple conformation system

In 2019, 1,1'-(ethane-1,2-diyl)dibenzene-2,2'-bis(methylene) (EDB) bridged glycosyl fluoride **6** was developed (Fig. 11).[25] Suzuki glycosylation[23] allowed the kinetic reaction of *o*-xylylene bridged **5** with high α-selectivity. The *o*-xylylene bridge, however, locked the pyranose ring into the $^3S_1$ conformation, which directed the 2-*O*-benzyl group axially toward

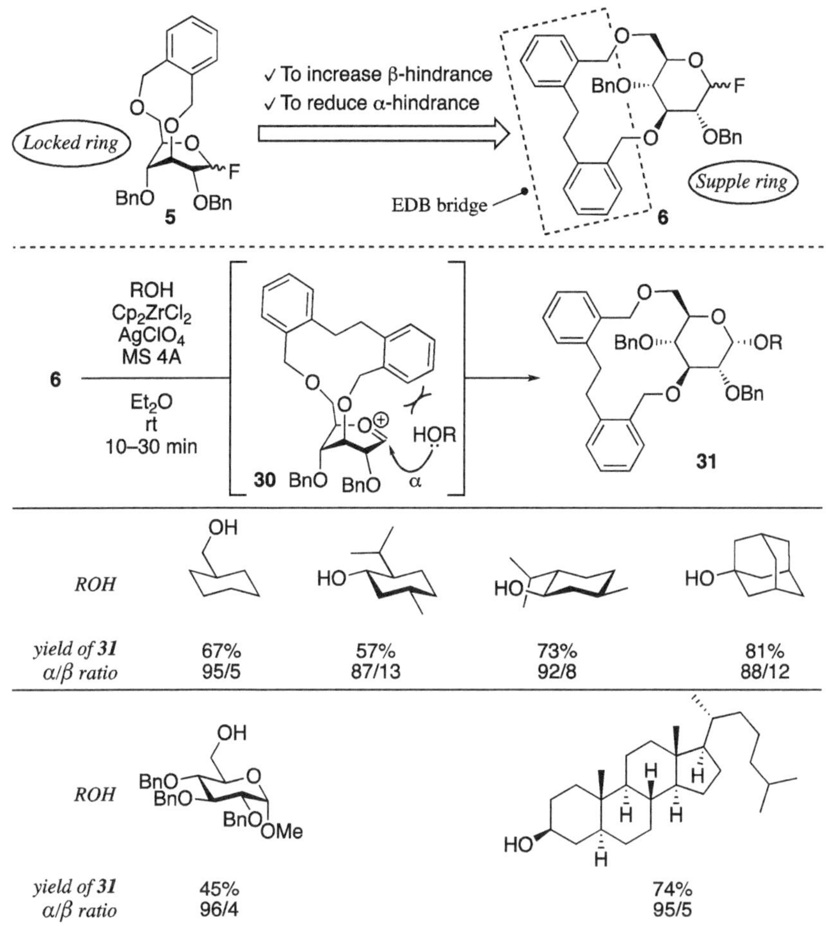

Fig. 11 Development of a supple conformation system by protection with the EDB bridging group between O-3 and O-6 on the D-glucopyranose ring and α-selective glycosylation with various alcohols.

the α face, thus inducing adverse steric hindrance. The α-selectivity of glucosides **29** therefore decreased with an increase in the reaction temperature. This issue was resolved by increasing the steric hindrance at the β face and reducing the overhang of the 2-*O*-benzyl group on the α face. To satisfy both of these requirements, the EDB group as a bridging group was developed, which had higher flexibility than the *o*-xylylene group. The glycosylation reaction using EDB-bridged D-glucosyl fluoride **6** afforded the corresponding glucoside **31** with a good α-selectivity even at room temperature within 30 min. The reaction proceeded through a corresponding oxocarbenium ion intermediate **30**, and similar α-selectivity was observed for **6**-α and **6**-β. A brief scope of the glycosylation reaction using **6** is summarized in Fig. 11.

The conformation of the glucopyranose ring bridged by the EDB group was converted from the $^{4}C_{1}$ conformation to various other conformations (Fig. 12).[25] The following method was developed to determine the steric conformations of the EDB-bridged glucopyranose rings. The coupling constants were obtained from the ^{1}H NMR spectrum (Step 1), and the dihedral

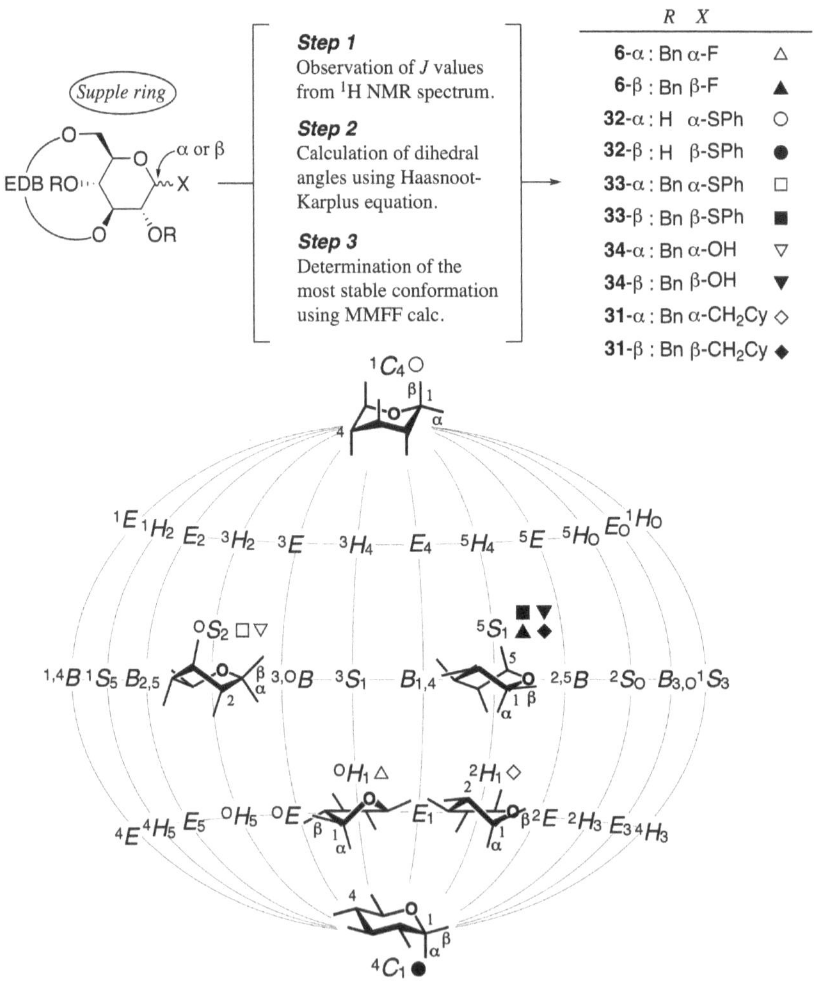

Fig. 12 Method used to determine the conformations of D-glucopyranose ring and resulting distribution map.

angles were calculated using the Haasnoot–Karplus equation (Step 2).[26] Because one coupling constant gave up to four solutions using this equation, the most stable conformation among the available combinations of dihedral angles was determined from using the Merck molecular force field (MMFF) calculations (Step 3). By analyzing the obtained conformations, **32**-α (R=H and X=α-SPh) adopted a $^{1}C_4$ conformation, whereas its anomeric isomer, **32**-β, possessed the same chair type but with a $^{4}C_1$ conformation. For **33**-α and **33**-β, where the two hydroxy groups were protected using benzyl groups (R=Bn), $^{O}S_2$ and $^{5}S_1$ conformations were assigned as a skew-boat type. As displayed on a map of conformations that put the $^{1}C_4$ and $^{4}C_1$ conformations on both poles (Cremer–Pople–Stoddart coordinates),[27,28] the calculated conformations of the glucopyranose system in **6** and **31**–**34** were widely distributed. This phenomenon suggested the attachment of the EDB group produced a bicyclic skeleton in which the newly formed ring modulated the conformation of the glucopyranose, competing with the glucopyranose ring's tendency to adopt the $^{4}C_1$ conformation.

The pyranose ring "supple" by the EDB bridging group in combination with the α-selective glycosylation reaction enabled the synthesis of small cyclodextrins (Fig. 13).[25] Cyclodextrins (CDs, IUPAC names: cyclomalto-oligosaccharides) are oligomers of cyclized α-(1→4)-D-glucopyranosides,

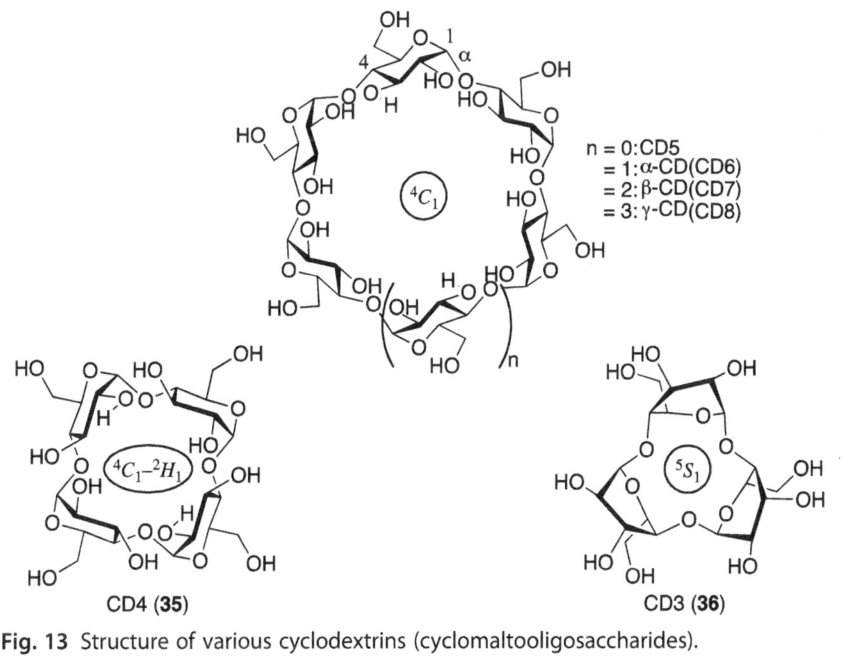

Fig. 13 Structure of various cyclodextrins (cyclomaltooligosaccharides).

mainly known as six- to eight-mers (CD6–CD8).[29] Because CD6–CD8 are non-toxic due to their being composed of only D-glucose,[30–32] and inexpensive owing to their mass production by enzymatic processes,[33] they are widely used not only in science research but also in consumer products, mainly for their inclusion (host–guest) abilities. Because the type of molecules capable of inclusion depends on the CD's vacant volume, CDs have been studied with respect to their size. The structure of the largest CD has been determined to be CD35,[34] and the smallest had been CD5.[35] The existence of smaller CD4 (**35**) and CD3 (**36**) had been questioned due to the difficulty in maintaining the original stable conformation of the α-(1→4)-D-glucopyranose.

The intramolecular glycosylation of the linear tetramer gave CD4 (Fig. 14).[25] The glycosylation using **37** and **38**, followed by removal of an acetyl group, provided dimer **39** with excellent α-selectivity. Higher selectivity was observed using **38** compared to using **6**. Because of changing the protecting group at O-2 from the benzyl group to a smaller allyl group, the steric hindrance reduced at the α face of **38**. Removal of the EDB groups on **38** provided the corresponding disaccharide whose pyranoses adopt $^4C_1$ conformations, and the coupling constant observed in the ^{1}H NMR spectrum confirmed the α stereochemistry on **39**. The glycosylation of **39** with **40** produced by the similar glycosylation from **37** and **38** followed by fluorination to afford tetramer **41** in an 82% yield. The stereochemistry of the newly formed glycosidic bond in **41** was undetermined because of the overlapping ^{1}H NMR signals. Subsequently, removal of the acetyl group at O-4 followed by an intramolecular glycosylation reaction, gave cyclic tetramer **42**. The ^{1}H and ^{13}C NMR spectra of cyclized **42** were in agreement with the pattern of a monosaccharide, indicating the unified stereochemistry of the four anomeric positions. Since α-stereochemistry at **39** and **40** was confirmed, anomeric linkages were all α. The reproducibility and yield of the intramolecular glycosylation were poor (six successes to 17 failures, ≤25% yield from **41**). Removal of the allyl groups by a combination of tetrakis(triphenylphosphine)palladium(0) and 1,3-dimethylbarbituric acid (**43**)[36] from the cyclic tetramer **42**, followed by hydrogenolysis, allowed the successful removal of the EDB group, which achieved the synthesis of CD4 (**35**). The ^{1}H NMR spectra revealed that the pyranose ring conformations were determined to be between $^4C_1$ and $^2H_1$ for CD4 (**35**) and $^2H_1$ for **42**. The pyranose moiety in these compounds was distorted and flatter than that observed in larger CDs.

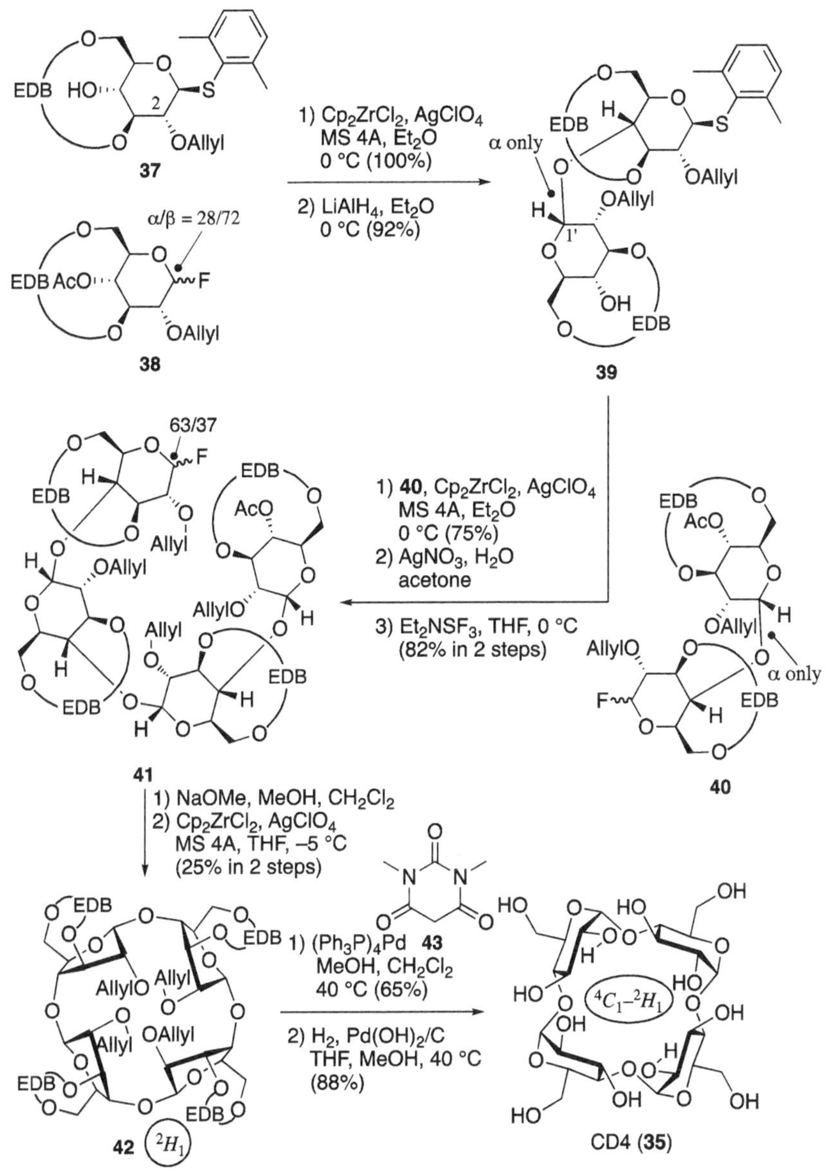

Fig. 14 Synthesis of CD4 using an EDB-bridging glycosyl donor by a supple conformation system.

A similar process to that used to prepare CD4 led to the synthesis of CD3 (Fig. 15).[25] The cyclization of the linear trimer **44** derived from dimer **39** and monomer **38** provided cyclic trimer **45**. The intramolecular glycosylation was more efficient than that with **41**, with the yield of product **45**

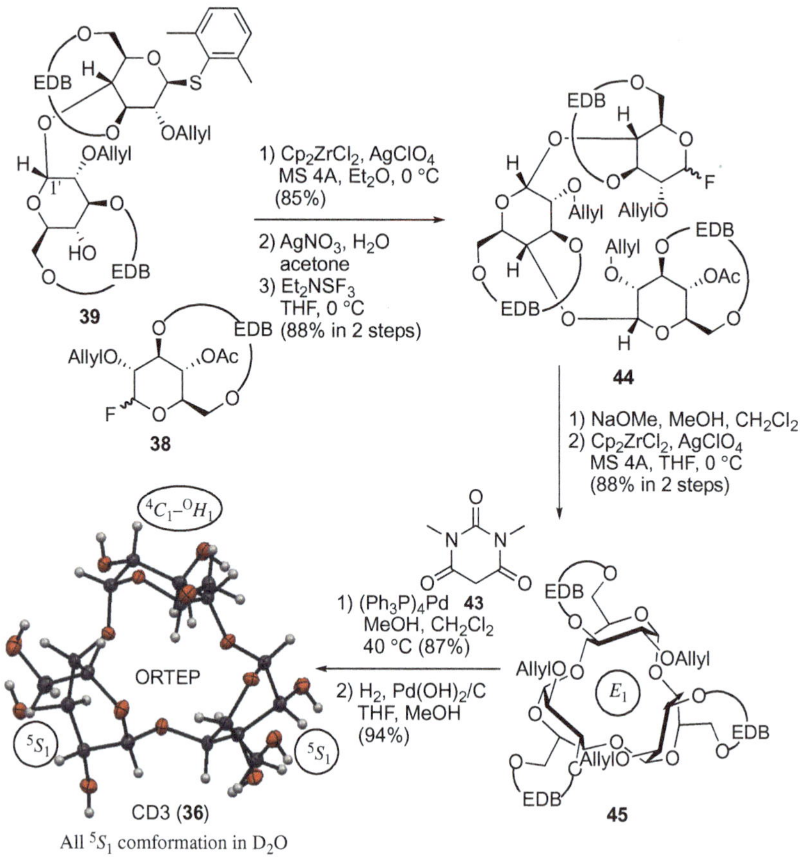

Fig. 15 Synthesis of CD3 using EDB-bridging glycosyl donor by a supple conformation system.

reaching an 88% yield. Removal of the allyl and EDB groups led to the formation of CD3 (**36**). ^{1}H and ^{13}C NMR spectroscopy and single-crystal X-ray diffraction studies confirmed the structure of CD3 (**36**). The NMR spectra of **45** and CD3 (**36**) were consistent with equivalent sugar units, and the average pyranose conformations were determined to be the E_1 and $^5S_1$ forms, respectively. The single-crystal X-ray diffraction analysis of CD3 (**36**) indicated that the conformations of the three pyranoses were different (two with $^5S_1$ and one between $^4C_1$ and $^OH_1$ conformations) in the crystal lattice.

5. Conclusions

Professor Yamada's 21 years of research at Kwansei Gakuin University led to the development of stereoselective glycosylation reactions focusing on the steric conformation of sugars. His consistent research led to the remarkable achievement of proving the existence of the smallest cyclodextrin that could ever exist. The supple conformation system proposed by Professor Yamada is a completely original concept, which will be widely applied to the synthesis of natural products and the creation of new functional materials, and will stand as a lasting and powerful contribution to carbohydrate chemistry. I express my regret on his untimely passing in the midst of his active career. I hope to preserve his unflagging respect for chemistry and shall always remember his sincere attitude toward research and his passion towards everything that he did.

References

1. Yamada, H.; Nishizawa, M. Synthesis and Structure Revision of Intensely Sweet Saponin Osladin. *J. Org. Chem.* **1995**, *60*, 386–397. https://doi.org/10.1021/jo00107a018.
2. Nishizawa, M.; Imagawa, H.; Kan, Y.; Yamada, H. Total Synthesis of *cyclo*-L-Rhamnohexaose by a Stereoselective Thermal Glycosylation. *Tetrahedron Lett.* **1991**, *32*, 5551–5554. https://doi.org/10.1016/0040-4039(91)80081-G.
3. Yamada, H.; Ikeda, T. Rhamnosylation Reaction With a Phenyl 1-Thiorhamnoside and a Rhamnosyl Fluoride Which Have $^4C_1$ Conformation. *Chem. Lett.* **2000**, *4*, 432–433. https://doi.org/10.1246/cl.2000.432.
4. Ikeda, T.; Yamada, H. β-Selective *O*-Rhamnosylation With a Rhamnosyl Trichloroacetimidate That Has the $^4C_1$ Conformation. *Carbohydr. Res.* **2000**, *329*, 889–893. https://doi.org/10.1016/S0008-6215(00)00240-8.
5. Yamada, H.; Tanigakiuchi, K.; Nagao, K.; Okajima, K.; Mukae, T. Ring Conformations of D-Glucose Derivatives Possessing Two Bulky Silyl Protecting Groups at the 3,4-Positions; The First Observation of a Stable Full-Axial Chair Conformer Without Bridge Structures. *Tetrahedron Lett.* **2004**, *45*, 5615–5618. https://doi.org/10.1016/j.tetlet.2004.05.126.
6. Yamada, H.; Tanigakiuchi, K.; Nagao, K.; Okajima, K.; Mukae, T. The First Ring Inversion of Pyranoses Induced by Bulky Silyl Protections at the 2- and 3-Positions. *Tetrahedron Lett.* **2004**, *45*, 9207–9209. https://doi.org/10.1016/j.tetlet.2004.10.078.
7. Okada, Y.; Mukae, T.; Okajima, K.; Taira, M.; Fujita, M.; Yamada, H. Highly β-Selective O-Glucosidation Due to the Restricted Twist-Boat Conformation. *Org. Lett.* **2007**, *9*, 1573–1576. https://doi.org/10.1021/ol070427b.
8. Masamune, S.; Choy, W.; Petersen, J. S.; Sita, L. R. Double Asymmetric Synthesis and a New Strategy for Stereochemical Control in Organic Synthesis. *Angew. Chem. Int. Ed. Engl.* **1985**, *24*, 1–30. https://doi.org/10.1002/anie.198500013.

9. Spijker, N. M.; van Boeckel, C. A. A. Double Stereodifferentiation in Carbohydrate Coupling Reactions: the Mismatched Interaction of Donor and Acceptor as an Unprecedented Factor Governing the α/β Ratio of Glycoside Formation. *Angew. Chem., Int. Ed. Eng.* **1991**, *30*, 180–183. https://doi.org/10.1002/anie.199101801.
10. Okada, Y.; Nagata, O.; Taira, M.; Yamada, H. Highly β-Selective and Direct Formation of 2-*O*-Glycosylated Glucosides by Ring Restriction into Twist-Boat. *Org. Lett.* **2007**, *9*, 2755–2758. https://doi.org/10.1021/ol070720b.-.
11. Kasai, Y.; Michihata, N.; Nishimura, H.; Hirokane, T.; Yamada, H. Total Synthesis of (+)-Davidiin. *Angew. Chem., Int. Ed.* **2012**, *51*, 8026–8029. https://doi.org/10.1002/anie.201203305.
12. Haddock, E. A.; Gupta, R. K.; Al-Shafi, S. M. K.; Haslam, E.; Magnolato, D. The Metabolism of Gallic Acid and Hexahydroxydiphenic Acid in Plants. Part 1. Introduction. Naturally Occurring Galloyl Esters. *J. Chem. Soc. Perkin Trans. 1* **1982**, *2515–2524*. https://doi.org/10.1039/P19820002515.
13. Spencer, C. M.; Cai, Y.; Martin, R.; Lilley, T. H.; Haslam, E. The Metabolism of Gallic Acid and Hexahydroxydiophenic Acid in Higher Plants Part 4; Polyphenol Interactions Part 3. Spectroscopic and Physical Properties of Esters of Gallic Acid and (*S*)-Hexahydroxydiphenic Acid With D-Glucopyranose ($^4C_1$). *J. Chem. Soc. Perkin Trans. 2* **1990**, *651–660*. https://doi.org/10.1039/p29900000651.
14. Zhu, M.; Phillipson, J. D.; Greengrass, P. M.; Bowery, N. E.; Ya, C. Plant Polyphenols: Biologically Active Compounds or Non-Selective Binders to Protein? *Phytochemistry* **1997**, *44*, 441–447. https://doi.org/10.1016/S0031-9422(96)00598-5.
15. Yamada, H.; Nagao, K.; Dokei, K.; Kasai, Y.; Michihata, N. Total Synthesis of (−)-Corilagin. *J. Am. Chem. Soc.* **2008**, *130*, 7566–7567. https://doi.org/10.1021/ja803111z.
16. Hatano, T.; Hattori, S.; Ikeda, Y.; Okuda, T.; Shingu, T. Galiotannins Having a 1,5-Anhydro-D-glucitol Core and Some Ellagitannins From Acer Species. *Chem. Pharm. Bull.* **1990**, *38*, 1902–1905. https://doi.org/10.1248/cpb.38.1902.
17. Okada, Y.; Asakura, N.; Bando, M.; Ashikaga, Y.; Yamada, H. Completely β-Selective Glycosylation Using 3,6-*O*-(*o*-Xylylene)-Bridged Axial-Rich Glucosyl Fluoride. *J. Am. Chem. Soc.* **2012**, *134*, 6940–6943. https://doi.org/10.1021/ja301480g.
18. Greene, T. W.; Wuts, P. G. M. *Protective Groups in Organic Synthesis*, 4th ed.; John Wiley & Sons: New York, 1999; pp. 166–171.
19. Schmidt, O. T.; Lademann, R. Corilagin, ein weiterer kristallisierter Gerbstoff aus Dividivi. X. Mitteilung über natürliche Gerbstoffe. *Justus Liebigs Ann. Chem.* **1951**, *571*, 232–237. https://doi.org/10.1002/jlac.19515710305.
20. Schmidt, O. T.; Schmidt, D. M.; Herok, J. Die Konstitution und Konfiguration des Corilagins. XIX. Mitteilung über natürliche Gerbstoffe. *Justus Liebigs Ann. Chem.* **1954**, *587*, 67–74. https://doi.org/10.1002/jlac.19545870107.
21. Rubio, E. M.; García-Moreno, M. I.; Balbuena, P.; Mellet, C. O.; García Fernández, J. M. Rigid Spacer-Mediated Synthesis of Bis-Spiroketal Ring Systems: Stereoselective Synthesis of Nonsymmetrical Spiro Disaccharides. *Org. Lett.* **2005**, 7, 729–731. https://doi.org/10.1021/ol0474094.
22. Mukaiyama, T.; Maeshima, H.; Jona, H. Catalytic and Stereoselective Glycosylation With Disarmed Glycosyl Fluoride by Using a Combination of Stannous(II) Chloride ($SnCl_2$) and Silver Ttetrakis(pentafluorophenyl)borate [$AgB(C_6F_5)_4$] as a Catalyst. *Chem. Lett.* **2001**, *30*, 388–389. https://doi.org/10.1246/cl.2001.388.
23. Suzuki, K.; Maeta, H.; Matsumoto, T.; Tsuchihashi, G.-I. New Glycosidation Reaction 2. Preparation of 1-Fluoro-D-desosamine Derivative and its Efficient Glycosidation by the Use of Cp_2HfCl_2–$AgClO_4$ as the Activator. *Tetrahedron Lett.* **1988**, *29*, 3571–3574. https://doi.org/10.1016/0040-4039(88)85295-X.

24. Motoyama, A.; Arai, T.; Ikeuchi, K.; Aki, K.; Wakamori, S.; Yamada, H. α-Selective Slycosylation of 3,6-*O*-*o*-Xylylene-Bridged Glucosyl Fluoride. *Synthesis* **2018**, *50*, 282–294. https://doi.org/10.1055/s-0036-1590927. Erratum; *Synthesis* **2018**, *50*, 4695. DOI: 10.1055/s-0037-1610841.
25. Ikuta, D.; Hirata, Y.; Wakamori, S.; Shimada, H.; Tomabechi, Y.; Kawasaki, Y.; Ikeuchi, K.; Hagimori, T.; Matsumoto, S.; Yamada, H. Conformationally Supple Glucose Monomers Enable Synthesis of the Smallest Cyclodextrins. *Science* **2019**, *364*, 674–677. https://doi.org/10.1126/science.aaw3053.
26. Haasnoot, C. A. G.; de Leeuw, F. A. A. M.; Altona, C. The Relationship Between Proton–Proton NMR Coupling Constants and Substituent Electronegativities-I. An Empirical Generalization of the Karplus Equation. *Tetrahedron* **1980**, *36*, 2783–2792. https://doi.org/10.1016/0040-4020(80)80155-4.
27. Cremer, D.; Pople, J. A. A General Definition of Ring Puckering Coordinates. *J. Am. Chem. Soc.* **1975**, *97*, 1354–1358. https://doi.org/10.1021/ja00839a011.
28. Stoddart, J. F. *Stereochemistry of Carbohydrates*; Wiley: New York, 1971; pp. 50–58.
29. Szejtli, J. Introduction and General Overview of Cyclodextrin Chemistry. *Chem. Rev.* **1998**, *98*, 1743–1753. https://doi.org/10.1021/cr970022c.
30. World Health Organization. Evaluation of Certain Food Additives: α-Cyclodextrin. *WHO Tech. Rep. Ser.* **2005**, *928*, 16–20.
31. World Health Organization. Evaluation of Certain Food Additives and Contaminants: β-Cyclodextrin. *WHO Tech. Rep. Ser.* **1995**, *859*, 28–29.
32. World Health Organization. Evaluation of Certain Food Additives and Contaminants: γ-Cyclodextrin. *WHO Tech. Rep. Ser.* **2000**, *896*, 26–27.
33. Szejtli, J. Past, Present, and Future of Cyclodextrin Research. *Pure Appl. Chem.* **2004**, *76*, 1825–1845. https://doi.org/10.1351/pac200476101825.
34. Ueda, H.; Endo, T. Large Ring Cyclodextrins. In *Cyclodextrins and Their Complexes: Chemistry, Analytical Methods, Applications*; Dodziuk, H., Ed.; Wiley-VCH: New York, 2006; pp. 370–380.
35. Nakagawa, T.; Ueno, K.; Kashiwa, M.; Watanabe, J. The Stereoselective Synthesis of Cyclomaltopentaose. A Novel Cyclodextrin Homologue With D.P. Five. *Tetrahedron Lett.* **1994**, *35*, 1921–1924. https://doi.org/10.1016/S0040-4039(00)73196-0.
36. Tsukamoto, H.; Suzuki, T.; Kondo, Y. Remarkable Solvent Effect on Pd(0)-Catalyzed Deprotection of Allyl Ethers Using Barbituric Acid Derivatives: Application to Selective and Successive Removal of Allyl, Methallyl, and Prenyl Ethers. *Synlett* **2007**, 3131–3136. https://doi.org/10.1055/s-2007-992352.

CHAPTER THREE

Chemical synthesis of sialoglyco-architectures☆

Hiromune Ando[a,*], Naoko Komura[a], Hide-Nori Tanaka[a], Akihiro Imamura[a,b], and Hideharu Ishida[a,b]

[a]Institute for Glyco-Core Research (iGCORE), Gifu University, Gifu, Japan
[b]Department of Applied Bioorganic Chemistry, Faculty of Applied Biological Sciences, Gifu, Japan
*Corresponding author: e-mail address: hando@gifu-u.ac.jp

Contents

Abbreviations

Ac acetyl
Bn benzyl
Boc *tert*-butoxycarbonyl
Bz benzoyl
CAc chloroacetyl
Cbz benzyloxycarbonyl
Cer ceramide
DABCO 1,4-diazabicyclo[2.2.2]octane

☆Dedicated to the memory of Professor Hidetoshi Yamada.

Advances in Carbohydrate Chemistry and Biochemistry, Volume 81
ISSN 0065-2318
https://doi.org/10.1016/bs.accb.2022.09.003

DBU	1,8-diazabicyclo[5.4.0]undec-7-ene
DMF	*N,N*-dimethylformamide
DTBS	di-*tert*-butylsilylene
EDC	1-ethyl-3-(3-dimethylaminopropyl) carbodiimide
Gc	glycolyl (hydroxyacetyl)
HOBt	1-hydroxybenzotriazole
Isp	isopropylidene
MBn	*p*-methoxybenzyl
MP	*p*-methoxyphenyl
MS	molecular sieves
Neu	neuraminic acid (sialic acid)
NIS	*N*-iodosuccinimide
Piv	pivaloyl (*tert*-butylcarbonyl)
Pyr	pyridine
rt	room temperature
SE	2-(trimethylsilyl)ethyl
Su	succinimidyl
TBBz	*p*-*tert*-butylbenzoyl
TBP	*p*-*tert*-butylphenyl
^{t}Bu	*tert*-butyl
TFA	trifluoroacetyl
TfOH	trifluoromethanesulfonic acid
THF	tetrahydrofuran
TMS	trimethylsilyl
Troc	2,2,2-trichloroethoxycarbonyl

1. Introduction

Sialic acid (SA) is a unique monosaccharide found in both vertebrates and invertebrates and contains a carboxyl group at the anomeric position and a glycerol chain at the C-6 position. In addition, the C-3 position is deoxygenated, and the C-5 position features an amide group substituted for the usual hydroxyl moiety. The variety of functional groups arrayed in one hexagonal ring may account for the high degree of involvement of sialic acid-containing glycans (sialoglycans) in a broad range of biological processes mediated by glycan–protein and glycan–glycan interactions.[1] However, information on the roles of sialoglycans in these biological events remains limited, which impedes further progress in glycobiology and glycochemistry. As suggested by many researchers, the glycosidation of sialic acid has been centered on glycochemistry because of the considerable difficulty in its stereocontrol.[2] In the past 50 years of research on sialic acid chemistry, extensive efforts have been devoted to overcoming the issue of the unstable stereochemistry of the short-lived oxocarbenium ion of sialic acid. An important finding pioneered by Prof. Hasegawa, who was the first principal investigator of our group, was that the use of acetonitrile as a reaction medium assisted α-selective sialylation by promoting nucleophilic attack by the nitrile group on the β-face of the oxocarbenium ion. Since the publication of the first paper on acetonitrile-assisted sialylation in 1988,[3] our group searched for a robust method of sialylation that enables the comprehensive synthesis of sialoglycoconjugates with high stereoselectivity and efficiency. In this chapter, we describe our recent progress in the study of the sialylation method and its application in the synthesis of glycoconjugates and their functional probes for biological studies.

2. Sialyl donors for α-glycoside formation

2.1 *N*-Troc-sialyl donor

The acetonitrile-assisted method for selective sialylation facilitated the synthesis of sialoglycans owing to its simplicity and scalability, enabling the production of structurally defined sialoglycans and their analogs useful for biological studies. However, the high sensitivity of stereoselectivity to reaction conditions has limited the scope of its application. In addition, considering the structural diversity of sialic acid, the original method using an *N*-acetyl sialic acid donor was found to be inflexible owing to the difficulty in the conversion of the *N*-acetyl group into derivatives, such as *N*-glycolyl and *N*-glycosylglycolyl groups. To overcome this issue, we introduced a selectively removable 2,2,2-trichloroethoxycarbonyl group (Troc) at the C-5 amine, producing an *N*-Troc sialic acid donor.[4] Notably, installing a Troc group at the C-5 amino group significantly enhanced sialyl donor reactivity. The coupling reaction of the *N*-Troc sialyl donor **2** with the reactive diol acceptor in acetonitrile at −50 °C was completed within 20 min, yielding sialoside **4** in 54% yield (α/β = 4/1). In contrast, the corresponding classical *N*-Ac derivative (**1**) required 13 h at −25 °C to produce sialoside **3** in 23% yield as a single α isomer (Scheme 1A). In addition to increasing the reactivity, the Troc group improved the solubility of the reactant in the solvents used for glycosylation and facilitated the chromatographic separation of the sialylated product anomers by increasing the differences in mobility on silica gel. As expected, *N*-Troc sialosides are useful for the synthesis of various partially modified congeners (Scheme 1B).

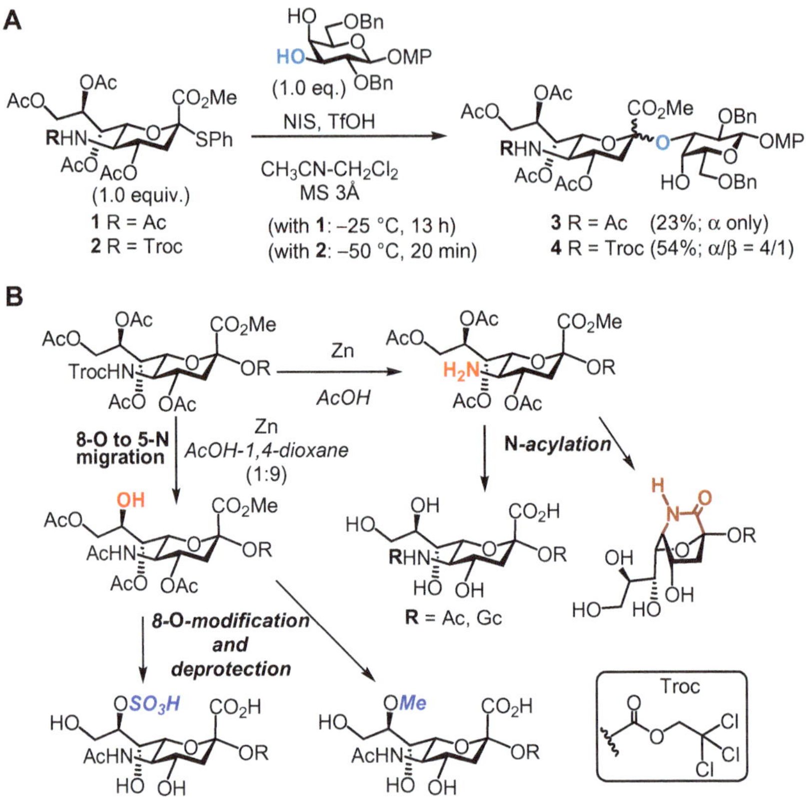

Scheme 1 Glycosidation of *N*-Troc sialyl donor (A) and conversion of *N*-Troc sialoside into various sialoside congeners (B).

The C-5-modified congeners Neu5Ac, Neu5Gc, and 1,5-lactamized-Neu can be derived by the selective deprotection of the Troc group using zinc in acetic acid followed by *N*-acylation. However, when the Troc group was removed under mild acidic conditions, the O-8 acetyl group migrated to the C-5 amino group, generating an 8-OH intermediate in a high yield. The 8-OH intermediate was used to synthesize 8-*O* partially modified congeners, such as 8-*O*-sulfonyl[5] and-methyl[6] congeners.

2.2 5*N*,7*O*-Oxazinanone sialyl donor

Considering the stereoselectivity of glycosidation, the *N*-Troc sialyl donor failed to provide consistent satisfactory results, which were affected by the structure of the coupling partners and the reaction conditions. To achieve

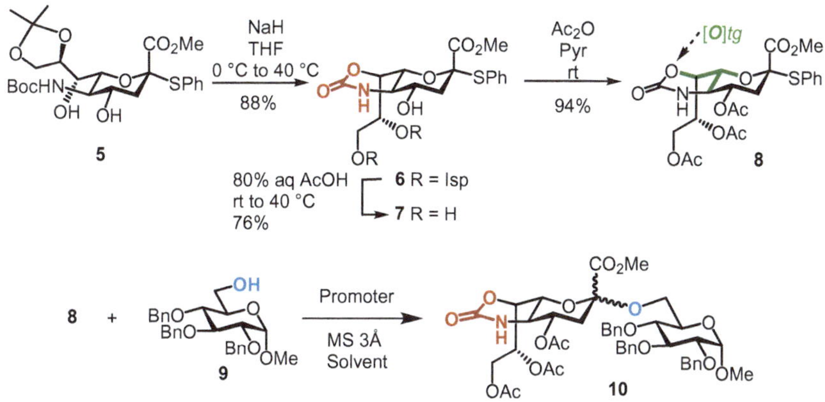

Entry	Promoter	Solvent	Temp (°C)	Time	Yield (%)	α/β
1	NIS (1.5 equiv) TfOH (0.3 equiv)	CH_3CN	–40	4 h	89	1/3.3
2	NIS (1.5 equiv) TfOH (0.3 equiv)	CH_2Cl_2	–40	4 h	70	1.1/1
3	PhSeBr (3.0 equiv) AgOTf (3.0 equiv)	CH_2Cl_2	–80 to rt	24 h	46	1.4/1
4	IBr (2.0 equiv) $AgClO_4{\bullet}H_2O$ (3.0 equiv)	CH_2Cl_2	–60	30 min	58	1.3/1

Scheme 2 Glycosidation of 5*N*,7*O*-oxazinanone sialyl donor.

the full stereocontrol of sialylation, we explored the modification of the sialyl donor. First, we introduced a cyclic carbamate at the C-5 and C-7 positions via the intramolecular transesterification of 5-NHBoc-sialic acid **5** with NaH, yielding 5*N*,7*O*-oxazinanone derivative **6**. Compound **6** was then transformed into the fully acetylated form **8** (Scheme 2).[7] A broad examination of the glycosylation of **8** revealed that glycosides were preferentially produced in most cases. Unexpectedly, the use of a nitrile solvent as the reaction medium enhanced the formation of β-glycosides. Furthermore, donor **8** was found to be less reactive and needed a strong activator, such as IBr-$AgClO_4{\bullet}H_2O$, to complete the glycosylation at low temperatures for a short time. The ring strain induced by the bicyclic system might have disfavored the formation of oxocarbenium cations from **8**. In addition, the conformation of the side chain of the 5*N*,7*O*-oxazinanone derivative **8** was fixed as the [*O*]*tg* conformation to the C-6—O-6 and C-5—C-6 bonds, which is inappropriate for stabilizing the oxocarbenium ion.[8] The glycoside preference in the glycosylation of **8** could be because of the steric hindrance of the axially oriented side chain. Similar results for the glycosidation of the 5*N*,7*O*-oxazinanone sialyl donor were separately reported by Crich and Wu.[9]

2.3 5-Ureido-sialyl donor

Next, we modified the 5-NH_2 group to the ureido form, yielding the sialyl donor **11**. A broad examination of the glycosidation revealed that **11** was apparently less reactive than the *N*-Troc sialyl donor **2** and fitted with a potent glycosylation-promoting reagent, IBr-AgOTf, to produce high yields of sialosides exhibiting excellent stereoselectivity in a short reaction time (Scheme 3).[10] The nucleophilic ureido moiety may capture a catalytic quantity of acid to slow the activation of thioglycoside in the case of NIS-TfOH-promoted glycosidation reaction. Therefore, the use of a stoichiometric quantity of acid in the promoting system, such as

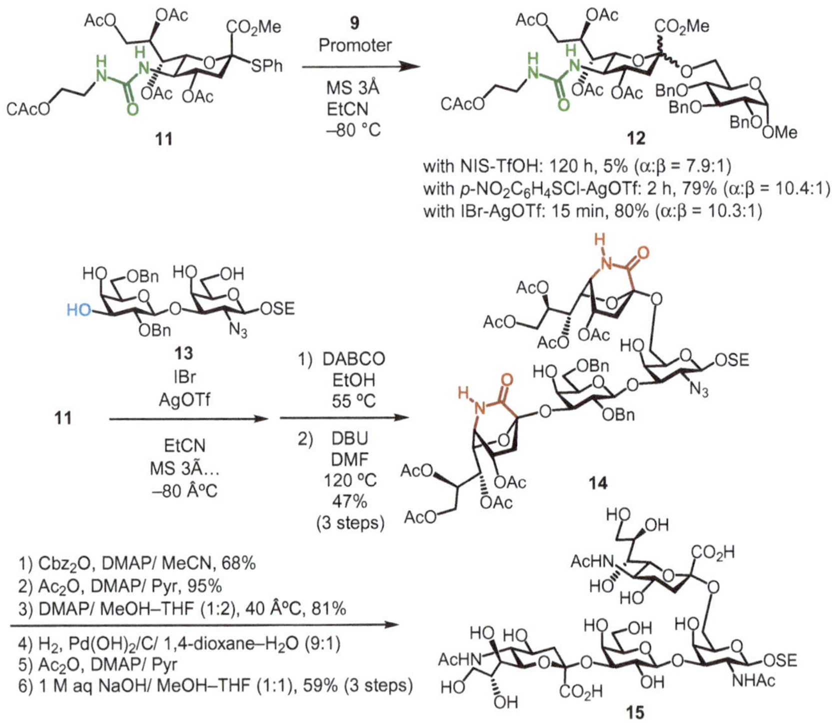

Scheme 3 Glycosidation of 5-ureido sialyl donor and the synthesis of Chol-1 ganglioside core **15**.

p-$NO_2C_6H_4SCl$-AgOTf or IBr-AgOTf, shortened the reaction time. We demonstrated that α-5-ureido-sialoside was selectively transformed into 1,5-lactamized sialoside via sequential basic treatment with DABCO and DBU in the presence of the β isomer. This selective transformation facilitated the chromatographic separation of the anomeric isomers and was successfully applied to the synthesis of a disialyl disaccharide, Chol-1 ganglioside core.[11] We sequentially performed the double sialylation of a tetraol derivative of Galβ(1,3) $GalN_3$ **13** with 5-ureido-sialyl donor **11** and 1,5-lactamization, which provided di-α-sialylated disaccharide **14** via easy chromatographic separation from β-sialylated byproducts. The lactamized sialic acid moieties were converted into their natural form as *N*-acetyl-sialic acid, delivering a Chol-1 ganglioside core **15**.

2.4 Bicyclic sialyl donor[12]

Concurrently with the above-detailed investigation of sialyl donors, we examined the possibility of the substrate-based stereocontrol of sialidation via bridgehead oxocarbenium ions. Inspired by the bridgehead carbocation (anti-Bredt carbocation) chemistry developed by Kraus et al,[13] we envisaged that an anti-Bredt oxocarbenium ion generated from a bicyclic α-sialic acid would restrict the attack of the nucleophile on the α-face via steric hindrance of the bicyclic system. We also anticipated that a bridgehead oxocarbenium ion with high ring strain would be more likely to form a tetrahedral sp^3 anomeric center with a glycosidic bond. The initial examination of the glycosidation reaction using macrocyclized sialyl donors with bicyclo [8.2.2] to [13.2.2] systems revealed that bicyclic sialyl donors were completely α-selective, and sialyl donor **20** with a bicyclo[12.2.2]system provided the highest yield of α-glycoside (83%) among the bicyclic donors tested. Although donor **21**, with one carbon longer macro-ring than **20**, appeared comparable to **20** in terms of reactivity, **21** was more amenable to 1,2-elimination, yielding a dehydro derivative as the major byproduct, which may stem from the higher tolerance of the more flexible **21** to a planar structure around the anomeric center as an sp^2 carbon atom. (Scheme 4)

Next, we introduced a 2,2-dichloroethoxycarbonyl moiety in the tether to enable the selective cleavage of the C-5 amine group and subsequent modification into natural congeners, yielding dichloro-sialyl donors **29** and **30**. These derivatives are glycosidated by a variety of hydroxylated molecules, including simple alcohols, monosaccharides, oligosaccharides,

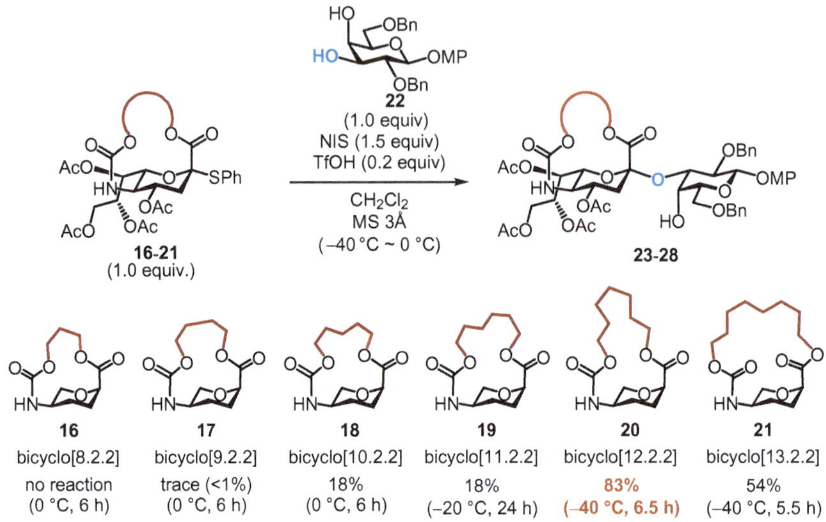

Scheme 4 Examination of glycosidation using bicyclic sialyl donors.

cholesterol, and polycyclic compounds. In all cases, sialosides were produced as single stereoisomers in high yields, indicating the broad substrate scope of the bicyclic sialyl donors. Phosphate **30** was also used to form *C*-glycoside bonds with carbon nucleophiles under acidic conditions (Scheme 5).

In comparing the prototype bicyclic donor **20** with the dichlorinated donor **29**, **29** was determined to be more reactive than **20** and was able to glycosidate at lower temperatures, such as −60 °C. This allowed the use of acid-labile protecting groups as coupling partners, thereby considerably expanding the applicability of the bicyclic sialyl donor. We assumed that the reactivity of the bicyclic donors was affected by the nucleophilicity of the C-5 carbamate moiety, which may inhibit the activation of the leaving groups by capturing the acid. Dichlorination of the tether may decrease the nucleophilicity of the carbamate moiety, inducing the dechlorinated donors to react much faster than the unsubstituted donor.

Importantly, the reactivity of the hydroxyl groups, such as those at the C-8 and C-4 positions of sialic acid, were enhanced in the bicyclic structure owing to the probable absence of hydrogen bonding between the C-5 amide moiety and the hydroxyl groups. The combined use of a bicyclic sialyl donor and acceptor with a free 8-OH enabled the synthesis of an α(2,8)-linked pentameric sialic acid in high yields (Scheme 6).

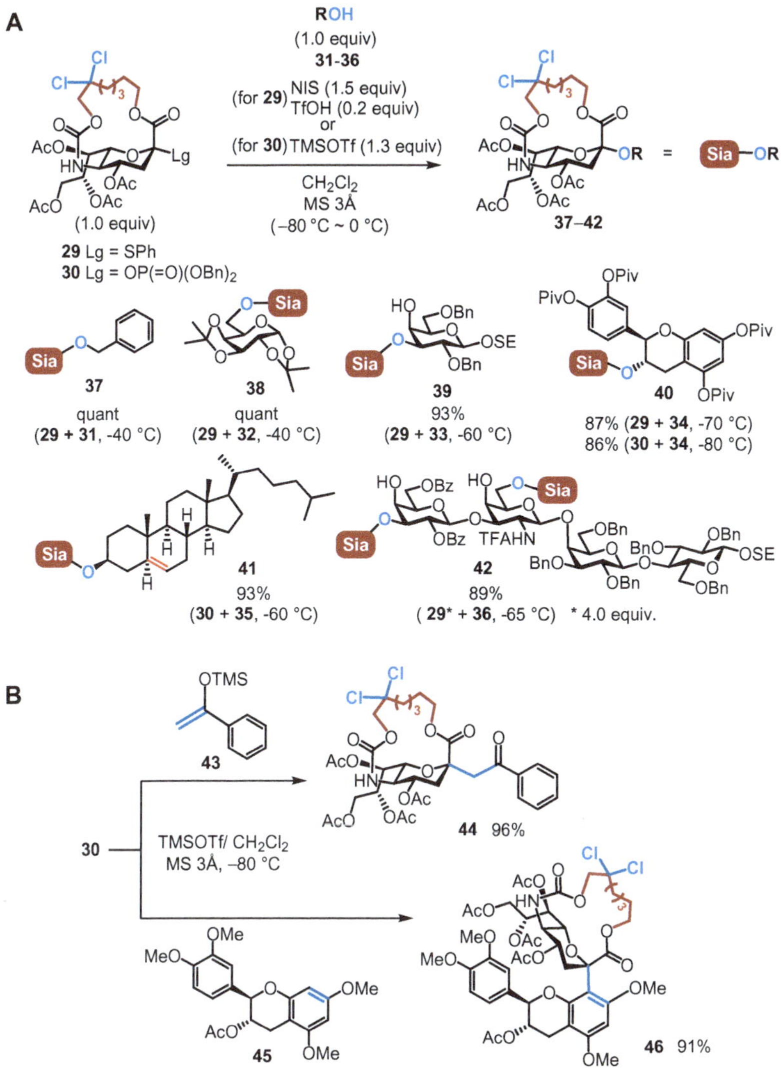

Scheme 5 Substrate scope study on fully α-selective glycosidation using bicyclic sialyl donors.

3. Ganglioside

Gangliosides are a large and diverse family of glycosphingolipids bearing a branched lipid called ceramide (Cer). In the exterior leaflet of the plasma membrane, gangliosides are anchored by their lipid moieties through

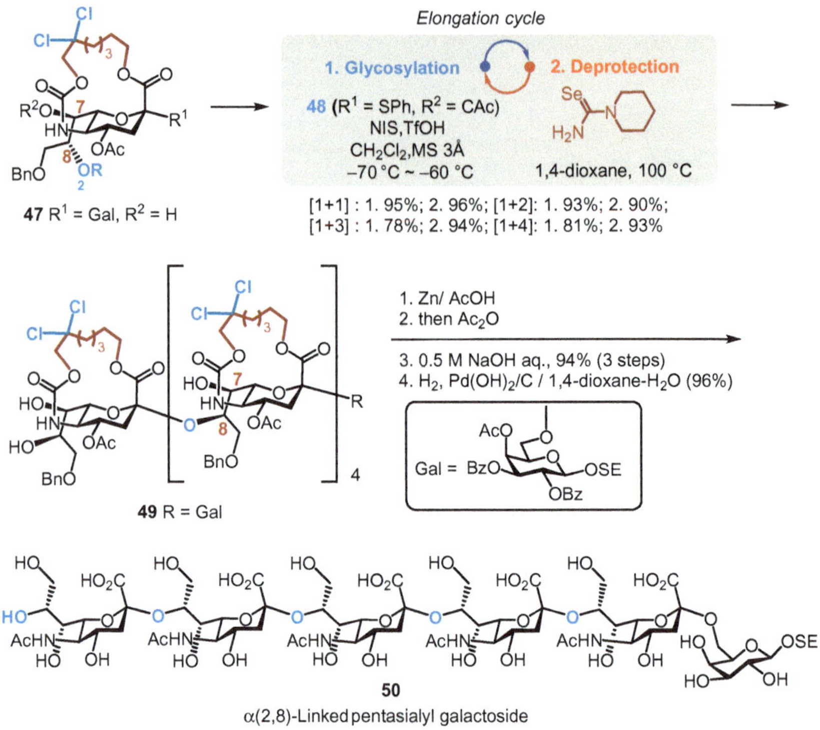

Scheme 6 Synthesis of an α(2,8)-linked pentameric sialoside using a bicyclic sialyl donor and acceptor.

hydrophobic interactions with other membrane molecules, thereby presenting their glycan moieties to the extracellular environment. Sialic acid exposes an array of polar functional groups to this exterior environment owing to its outermost positioning and its characteristic structure, which is comprised of an ω-amino acid-like moiety and a glycerol branch. Gangliosides in mammals are involved in crucial biological processes, including cell-type specific adhesion and the entry of toxins, viruses, and bacteria into cells via *trans*-interactions.[14] In addition, gangliosides modulate the function of protein receptors in the same plasma membrane via *cis*-interaction.[15] Significantly, gangliosides produced in echinoderms (e.g., starfish and sea cucumbers) are neurogenetically active toward model neuron cells (PC-12) in the presence of the neural growth factor.[16] Several echinodermatous gangliosides (EGs) show higher activity than the representative neuritogenic mammalian ganglioside GM1, making EGs appealing as potential precursors for neural disorder therapeutics. EGs differ from mammalian gangliosides and have unique glycan and Cer moieties. The glycan moieties of EGs contain partially modified sialic acid and oligomerized congeners. In addition, sialic acids, which have not been observed in mammalian gangliosides, are often observed as internal residues in EGs. More than 15 types of EGs have been structurally characterized, and their neuritogenic activity has been examined. However, the structure–activity relationships and molecular mechanisms of the potentiation of neurite generation by EGs are yet to be elucidated. To scrutinize the molecular basis of the diverse biological functions of gangliosides, the chemical synthesis of both their glycan moieties and their entire glycolipid structures is widely conducted. The creation of functional probes for gangliosides is equally important.

3.1 Complex mammalian ganglioside synthesis

3.1.1 Total synthesis of GalNAc-GD1a[17]

Ganglioside GalNAc-GD1a is highly localized in the ventral spinal roots of humans and bovines and in bovine spinal motor neurons. GalNAc-GD1a is closely associated with Guillain–Barré syndrome, which is the most frequent cause of acute flaccid paralysis. The ganglioside GalNAc-GD1a features a tandem branched sialyl disaccharide GM2-core. Although the branch that stems from the C-3 and C-4 positions of the galactose (Gal) residue is common in the ganglioside subfamily, its synthesis requires a careful selection of protecting groups for the C-2 and C-6 hydroxyl groups and of the sequence of glycosylation for the C-3 and C-4 hydroxyl groups. Contrary to conventional glycan synthesis, the glycosylation of the less reactive 4-OH should follow that of 3-OH with sialic acid owing to the low accessibility of the bulky sialic acid to the 3-OH. Significantly, an electron-donating protecting group should be added onto the 2-OH and 6-OH groups. Therefore, the construction of a sialyl α(2,3)-Gal derivative is crucial for synthesizing the GM2-core. We used *N*-Troc-sialyl Gal derivative **4**, which is suitable for large-scale production, as the key intermediate in the synthesis of the GM2-core (Scheme 7). The chemoselective removal of Troc was followed by acetylation to produce Neu-Gal acceptor **51**, which was then glycosylated with GalN donor **52** in a high yield, furnishing the GM2-core derivative. Next, the GM2-core was treated with zinc powder and acetic acid in DMF to deprotect the amino group on the GalN residue, and the acetyl group at the C-3 position subsequently migrated to the retrieved amine, yielding GM2-core acceptor **54**. The GM2-core derivative was also converted into a glycosyl imidate in six steps. The coupling reaction of

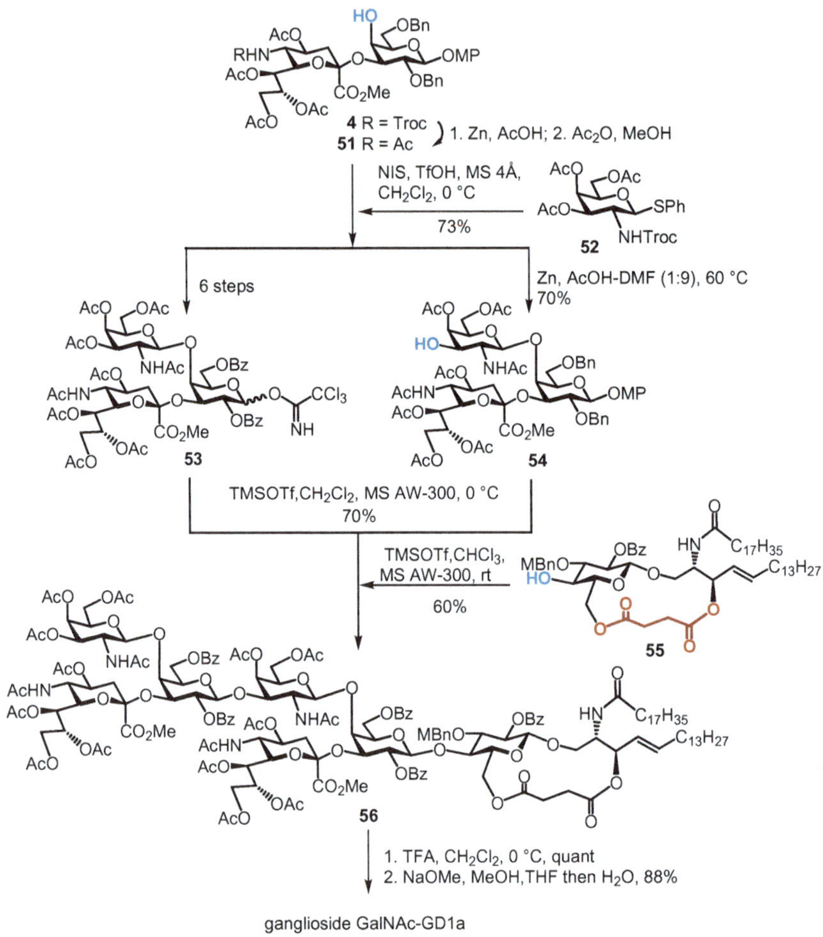

Scheme 7 Total synthesis of ganglioside GalNAcGD1a.

GM2-core donor **53** and acceptor **54** generated a tandem GM2-core in high yield. The hexasaccharide underwent protecting group manipulation and imidoylation to produce a hexasaccharyl donor. In the final glycosylation, cyclized Glc-Cer cassette **55** served as the coupling partner for the hexasaccharyl donor to furnish protected GalNAc-GD1a, which was then fully deprotected, delivering GalNAc-GD1a. By applying a similar method, analogous GalNAc-GM1b was synthesized efficiently.[18]

3.2 Synthesis of gangliosides bearing sialic acids as inner sugar residues

Mammalian gangliosides are highly organized and can be divided into several groups based on their glycan sequences. In contrast, EGs have less well-defined structures, some of which contain unique sialic acid linkages in their glycans. The most significant challenge in EG ganglioside synthesis is to construct sialic acid-embedded glycan structures, such as those involved in ganglioside GP3.[19] In GP3, a pair of Neu5Ac residues are flanked by α-galactofuranosyl (Galf) residues at the C-4 hydroxyl groups to be incorporated in the nonasaccharide moiety. The main challenge in ganglioside synthesis is the design of highly reactive sialic acid synthetic units as glycosyl donors and acceptors.

3.2.1 Total synthesis of GP3[20]

For the synthesis of the inner sialic acid residues in ganglioside GP3, which are connected to α-Gal residues at the 4-OH, the Troc modification of the C-5 amino group was used to achieve efficient glycosylation at the anomeric position and glycosylation at the 4-OH (Scheme 8). Among the possible *N*-Troc sialyl donors protected at the 4-OH with a selectively cleavable group, three derivatives were examined for glycosylation with a Gal acceptor. The results for glycosidation with Gal acceptor **59** showed that using 4,7,8-trichloroacetyl derivative **57** was the most effective means for obtaining α-sialoside and indicated a strong effect of glycerol modification on sialylation. *N*-Troc sialyl donor **57** also reacted with the Galf acceptor **58** to provide a high yield of α-sialoside. The two sialosides were converted into glycosyl acceptors **60** and **61** via the selective cleavage of the CAc groups by 1-selenocarbamoylpiperidine.[21] The *N*-Troc sialyl Gal acceptors were subjected to [2 + 2] coupling reactions. As anticipated, both hydroxyl groups were sufficiently reactive to be glycosylated with the disaccharyl donors, producing high yields of tetrasaccharides. The tetrasaccharides underwent protecting group manipulation, producing the tetrasaccharyl donors **62** and **63**. Subsequent [4 + 4] and [8 + Glc-Cer] coupling produced protected GP3, which was fully deprotected to produce ganglioside GP3.

3.3 Synthesis of disialic acid-containing structures

3.3.1 Sialyl acceptor for making disialic acids

Dimeric sialic acids (disialic acids), which are linked through α(2,4)-, α(2-8)-, α(2,9)-, or α(2,11)-glycosidic linkages, are characteristic of gangliosides. The construction of α(2,4)- and (2-8)-linkages is hampered by the low reactivity of the C-4 and C-8 hydroxyl groups of Neu5Ac because of hydrogen bonding with the adjacent amido group (Scheme 9). To overcome this issue, several research groups, including our group, have developed C-5-modified sialic acid derivatives as sialic acid acceptors.[22] We developed 1,5-lactamized sialic acid derivatives as glycosyl acceptors **67**

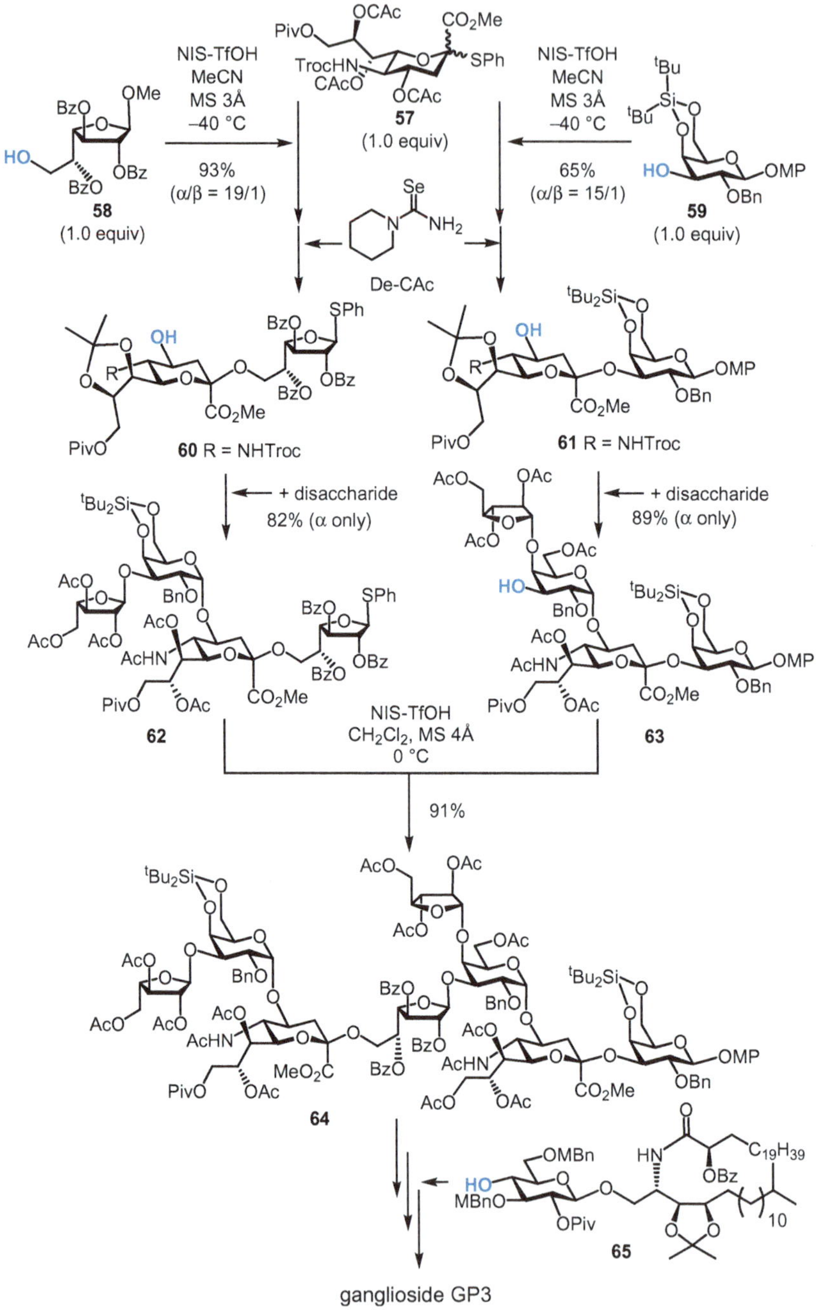

Scheme 8 Total synthesis of ganglioside GP3.

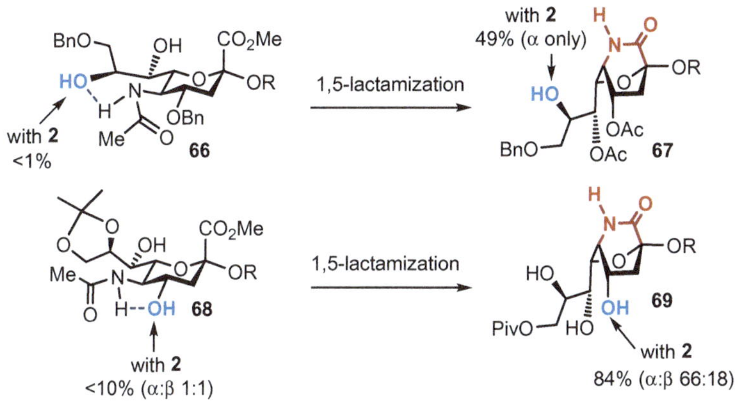

Scheme 9 1,5-Lactamization of sialic acid ameliorates the reactivity of 4-OH and 8-OH.

and **69**.[5] Glycosylation with sialyl donors and other glycosyl donors indicated that the 4-OH and 8-OH groups of the 1,5-lactamized sialyl acceptors were highly reactive and provided high yields of glycosylated products. Similarly, modifying the C-5 amine with trifluoroacetyl and Troc groups increases the reactivity of the C-4 or C-8 hydroxyl groups.

3.3.2 Total synthesis of HLG-2 [6a,23]

The 1,5-lactamized sialyl acceptor was used with the *N*-Troc sialyl donor in the total synthesis of ganglioside HLG-2, in which a tandem mixture composed of two sialic acid species, Neu5Gc and NeuAc, was incorporated into the glycan moiety (Scheme 10). To construct α(2,4)-linked disialic acid, the 1,5-lactam sialyl acceptor was investigated, and diol derivative **70** was identified as the most suitable. The glycosylation of 1,5-lactam sialyl acceptor **70** with *N*-Troc sialyl donor **2** produced the desired glycoside in a regio- and stereoselective manner. The Troc group was selectively cleaved by zinc in acetic acid, and the resulting amine was converted to acetoxyacetamide. Cbz was introduced at the lactam moiety, and the amide bond was hydrolyzed under basic conditions. Following protecting-group manipulation, the trisaccharyl donor was subjected to direct glycosylation with Cer to produce the HLG-2 framework, which then underwent global deprotection, providing ganglioside HLG-2.[23] The synergistic use of the *N*-Troc sialyl donor and 1,5-lactam sialyl acceptor also enabled the synthesis of Hp-s6 glycan, which has a partially modified α(2,8)-linked disialic acid residue, 8-sulfo-NeuAcα (2,8)Neu.[5]

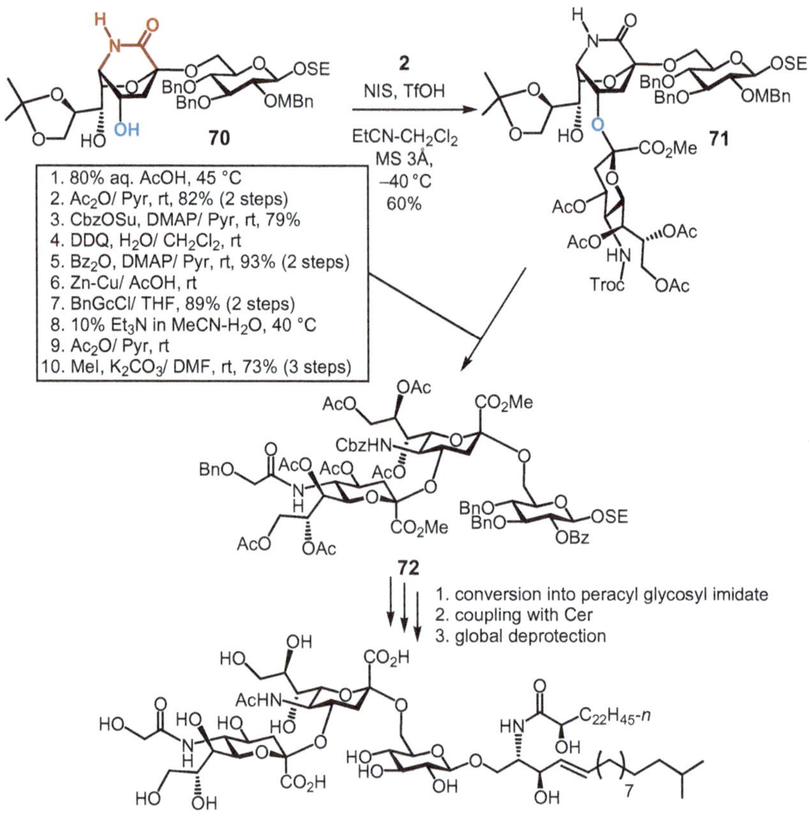

ganglioside HLG-2

Scheme 10 Total synthesis of ganglioside HLG-2.

3.3.3 Total synthesis of LLG-3 [6a]

Ganglioside LLG-3 is distinguished from other gangliosides by its unique glycan, which contains an 8-*O*—Me—NeuAcα(2,11)NeuGc sequence. *N*-Troc sialyl donor **2** was used as the key unit to access tandem sialic acid congeners (Scheme 11). *N*-Troc sialyl donor **2** was used to obtain Neu5NHTroc-Gal **4**, which was converted to Neu5NH$_2$ **74** through the selective removal of the Troc group. 8-*O*-Methyl-Neu glycolate derivative **76** was prepared from compound **2** in a short reaction sequence that included 8-*O* to 5-*N* migration of the acetyl group after the deprotection of the Troc group under mild acidic conditions. i.e., the key step for obtaining the 8-hydroxy intermediate. The intermediate was then methylated at the C-8 hydroxyl group. Next, carboxylic acid units **76** and **74** underwent a coupling reaction mediated by EDC·HCl and HOBt,

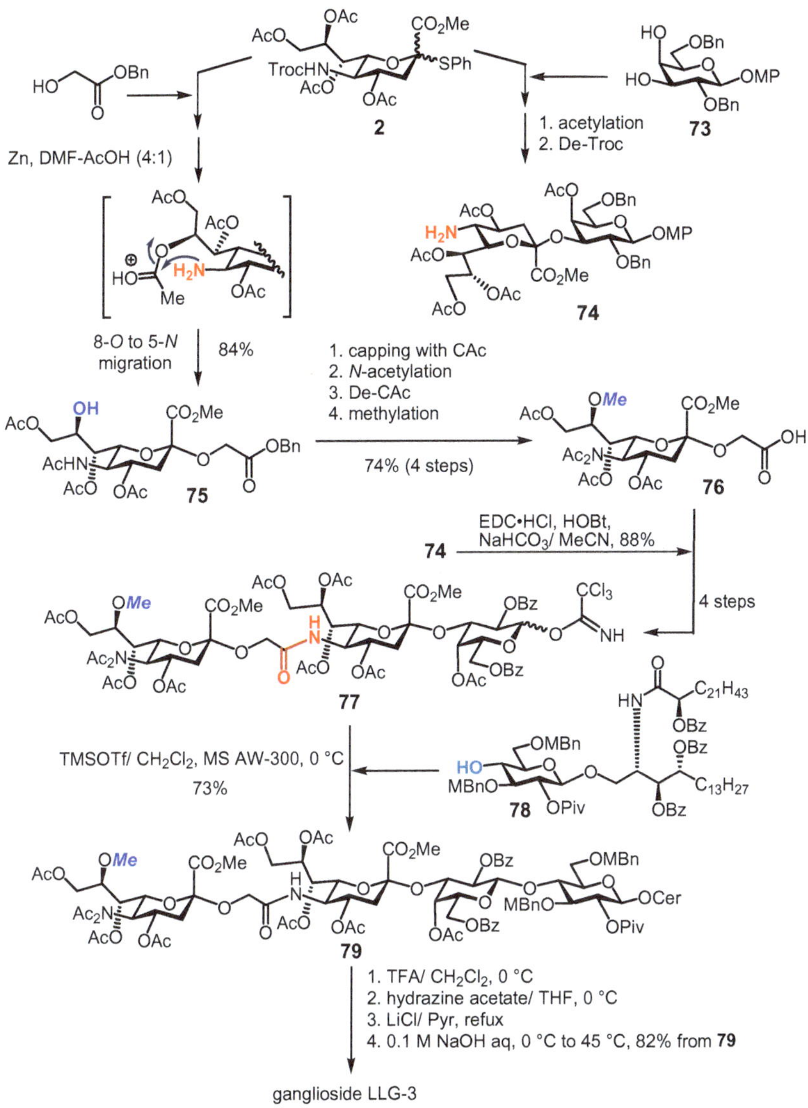

Scheme 11 Total synthesis of ganglioside LLG-3.

providing a high yield of trisaccharides. To synthesize the Cer moiety, the (*R*)-2-hydroxytricosanoic acid derivative was assembled from oxirane and icos-1-yne and coupled with phytosphingosine to yield phytoceramide. Phytoceramide was coupled with the Glc donor, followed by selective deprotection to produce Glc-Cer cassette **78**. Trisaccharide donor **77** was coupled with Glc-Cer cassette **78** to furnish the framework of ganglioside LLG-3 **79** in a high yield. Finally, compound **79** was deprotected to deliver ganglioside LLG-3.

4. Fluorescent probes for single molecule imaging

4.1 Gangliosides as the key molecules for the assembly and function of lipid raft

Gangliosides in the cell membrane participate in various cell-surface interactions as ligands for biomolecules such as proteins and glycoconjugates. Cell-surface interactions can be divided into *trans*- and *cis*-interactions. *Trans*-interactions are interactions between membrane molecules and exogenous molecules or membrane molecules residing in other cell membranes. *Cis*-interactions occur between molecules in the same cell membrane. In mammalian cells, gangliosides are involved in both types of interactions, functioning as receptors and modulators, mostly for proteins. In addition, gangliosides play a vital role in forming transient nanosized domains called lipid rafts,[24] which are important platforms for cell signaling and membrane trafficking. To understand such diverse ganglioside functions at the molecular level, the dynamic behavior of gangliosides in the cell membrane must be examined because all membrane molecules diffuse laterally during interactions. Single-molecule tracking has been used to investigate the interactions between membrane molecules and their counterparts.[25] To observe the behavior of gangliosides in the cell membrane via single-molecule tracking, fluorescent analogs of gangliosides that behave identically to native gangliosides in the cell membrane are required.

4.2 Chemical synthesis of fluorescently labeled gangliosides

Our method for synthesizing the entire structure of gangliosides was used to synthesize the fluorescently labeled gangliosides (Scheme 12).[26] The amino-modified analogs of gangliosides GM3 and GM1 were constructed with *N*-Troc-Neuα(2,3)Gal intermediate **4**, which was derivatized into various amino-modified synthetic units, such as compound **80** and Glc-Cer cassette **82**. Finally, the active esters of the fluorescent dyes were reacted with the amino gangliosides, producing fluorescent gangliosides. The key conversion of *N*-Troc-Neu-Gal into amino derivatives circumvented sialylation using an amino-modified sialyl donor or galactosyl acceptor,

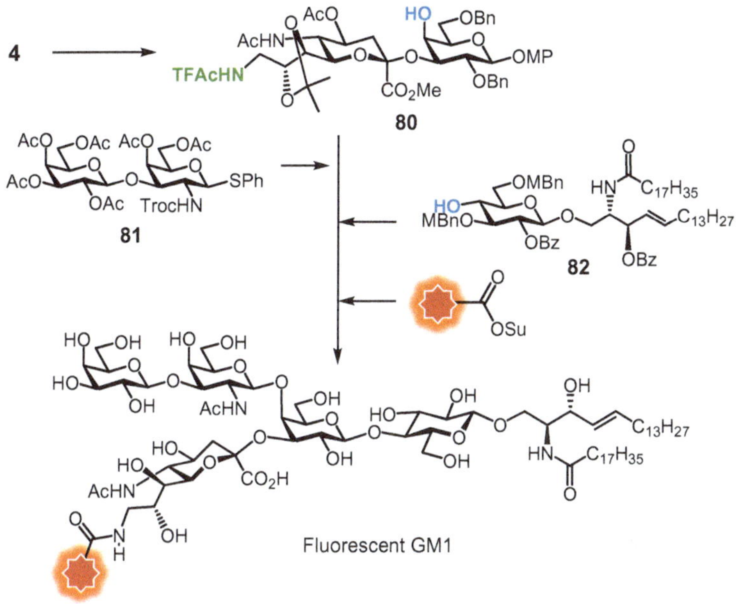

Scheme 12 Outline of the synthesis of fluorescently labeled ganglioside GM1.

whose stereoselectivity and product yield were strongly affected by the modification of the sialyl donor and its coupling partner.

For the synthesis of fluorescently labeled GQ1b with two disialic acid residues, we constructed a 9-amino modified disialic acid moiety by the 8-OH sialylation of a 1,5-lactamized sialyl galactosyl acceptor **83** in a high yield. The disialyl galactoside thus obtained was then converted into a glycosyl donor and successfully combined with a branched tetrasaccharide acceptor **85** to yield the core glycan moiety of GQ1b. Following our GlcCer cassette approach, the heptasaccharide moiety was extended to the full structure of GQ1b, which was finally conjugated with a fluorescent dye (Scheme 13).[27]

After fully α-selective sialylation became available in our group, a bicyclic 9-trifluoroacetamido sialyl donor **87** was developed and used for the direct sialylation of oligosaccharide moiety of SSEA-4, affording a sialyl oligosaccharide in high yield. Similarly to the fluorescent GM1 synthesis, the oligosaccharide was transformed into a fluorescent SSEA-4 probe (Scheme 14).[28]

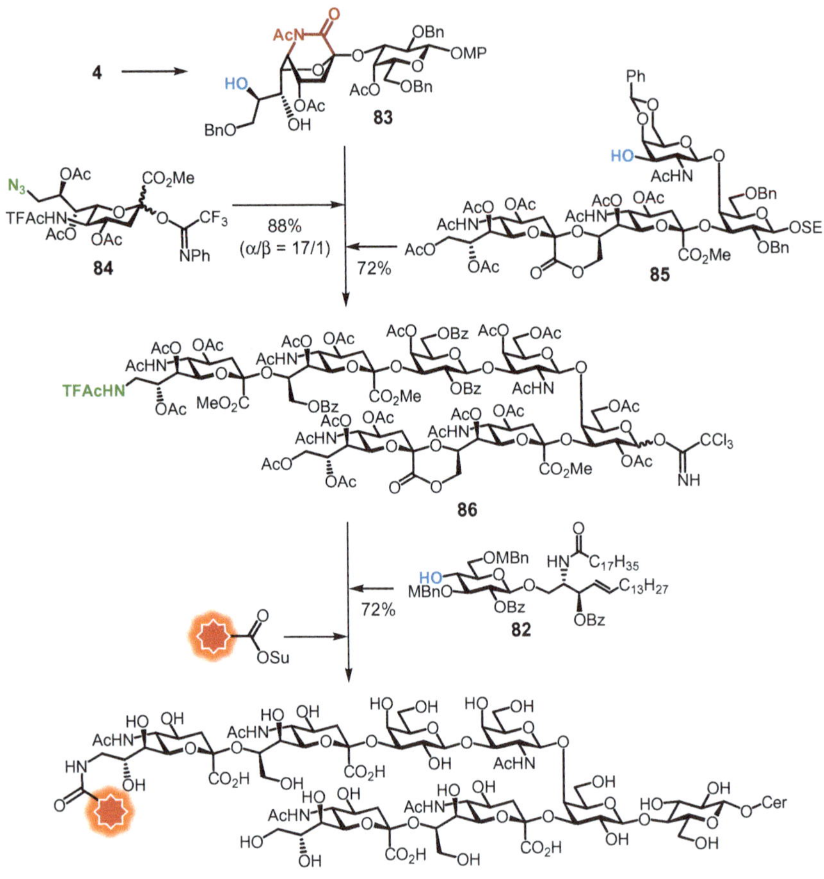

Scheme 13 Outline of the synthesis of fluorescently labeled ganglioside GQ1b.

4.3 Single molecule imaging of ganglioside in the live cell membrane

We examined whether the fluorescent molecules were biophysically equivalent to native gangliosides by partitioning fluorescent probes into detergent-resistant membranes and liquid-ordered, phase-like domains formed in giant plasma membrane vesicles. Incorporating hydrophilic fluorescent dyes (e.g., ATTO594, ATTO488, and fluorescein) at the outermost position of the glycan moiety has been demonstrated as the most suitable means for retaining the nature of gangliosides in the fluorescent analogs.[26a,c]

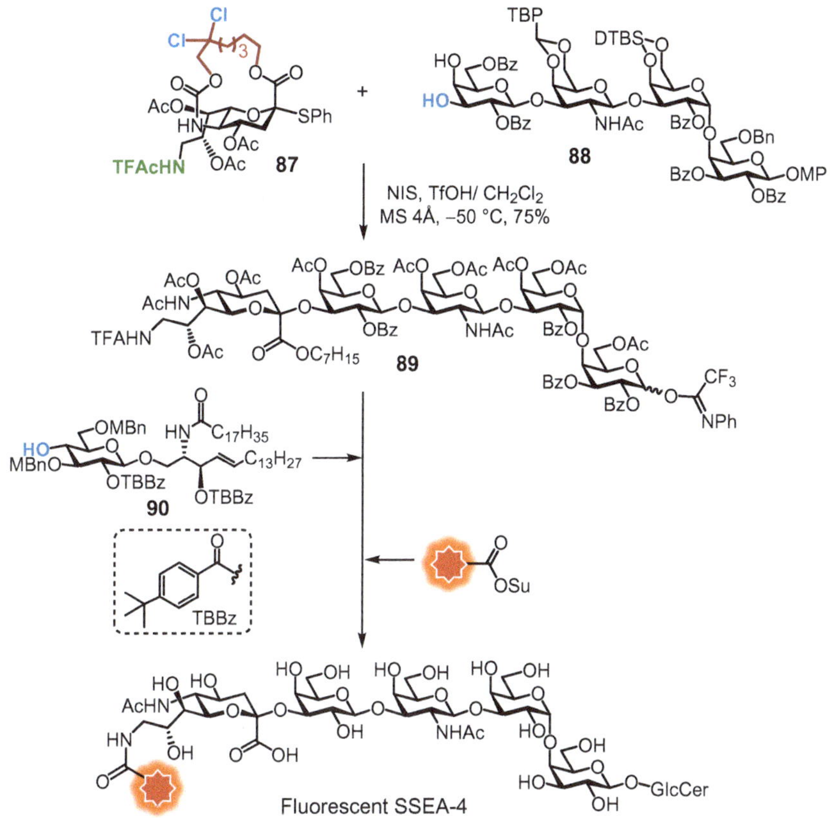

Scheme 14 Outline of the synthesis of fluorescently labeled ganglioside SSEA4 benefitting from the direct α-sialylation of an oligosaccharide acceptor with a bicyclic sialyl donor.

Single-molecule imaging of the interactions of the fluorescent ganglioside analogs with a fluorescent GPI-anchored protein, CD59, which forms the core for the formation of lipid raft domains, was performed in a live cell plasma membrane. Ganglioside GM3 transiently colocalized and co-diffused with CD59, forming CD59 clusters with a lifetime of 98ms in a cholesterol-dependent manner. In quiescent cells, gangliosides show free lateral diffusion.[26a] These results suggest that raft domains are dynamic entities in which gangliosides are dynamically exchanged between the raft and bulk domains (Fig. 1).

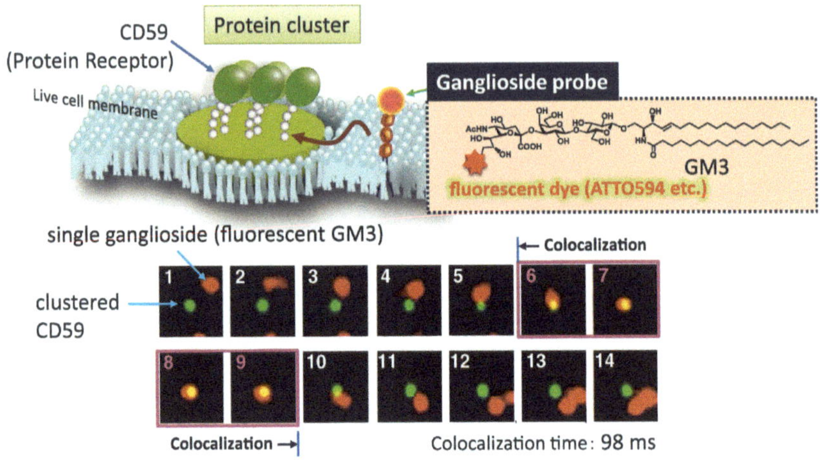

Fig. 1 Single molecule imaging of fluorescent GM3 probe in the live cell membrane.

5. Summary and outlook

Starting from the development of nitrile-assisted α-sialylation, we explored various sialyl donors, including *N*-Troc, 5*N*,7*O*-oxazinanone, and 5-ureido derivatives, which showed moderate to high α-selectivity. The *N*-Troc sialyl donor enabled the preparation of a sialyα(2,3)galactose unit, i.e., the essential common building block for the synthesis of mammalian gangliosides and their functionalized analogs for biological studies. Furthermore, the *N*-Troc sialyl donor enabled the synthesis of echinodermatous gangliosides containing a variety of sialic acid congeners. Notably, we developed macrobicyclic sialyl donors that ensured fully α-selective sialylation. The prominent features of this method, including the complete stereoselectivity, high glycosidation yield, broad substrate scope, and high applicability to sialic acid congeners, could revolutionize the synthesis of sialoglycans. Our method ensures the direct α-sialylation of oligosaccharides and subsequent diversification into sialic acid congeners, which has been neglected in most prior syntheses because of the extremely arduous separation of stereoisomers. Accordingly, the fundamental rule of the retrosynthesis of sialoglycans can be redefined in a straightforward manner. Furthermore, we envisage that integrating our method into the current advances in automated oligosaccharide synthesis would lead to the development of revolutionary technology.

Acknowledgments

We would like to thank all the past and current members of the Bioactive Molecule Science Laboratory of Gifu University for their contributions to the development of the sialoside syntheses described herein. We would like to thank Emeritus Professor Maoko Kiso and late Professor Akira Hasegawa, who pioneered sialic acid chemistry at Gifu University. We also acknowledge JSPS KAKENHI grants no. JP18H0392 (H.A.), JP19K23627 (N.K.), JP20K15412 (N.K.), JP18K05461 (H.-N.T.), JP18K05455 (A.I.), JSPS Core-to-Core Program (H.A.); grant No. JPJSCCA20200007 (H.A.), JST CREST (H.A.); grant No. JRMJCR18H2 (H.A.) and the Mizutani Foundation for Glycoscience (H.A.); a SUNBOR grant from the Suntory Foundation for Life Sciences (N.K.); grant No. JP20jk0210027 (A.I.) from Japan Agency for Medical Research and Development (AMED). We would like to thank Editage (www.editage.com) for the English language editing. Finally, we respectfully acknowledge the significant contribution of the late Professor Hidetoshi Yamada of Kwansei Gakuin University to carbohydrate chemistry and dedicate this article to the memory of his pioneering legacy.

References

1. (a) Edgar, L. J. Engineering the Sialome. *ACS Chem. Biol.* **2021**, *16*, 1829–1840; (b) Smith, B. A. H.; Bertozzi, C. R. The Clinical Impact of Glycobiology: Targeting Selectins, Siglecs and Mammalian Glycans. *Nat. Rev. Drug Discovery* **2021**, *20*, 217–243.
2. (a) De Meo, C.; Jones, B. T. Chemical Synthesis of Glycosides of *N*-Acetylneuraminic Acid. In *Advances in Carbohydrate Chemistry and Biochemistry*; Baker, D. C., Ed.; Vol. 75; Academic Press: New York, 2018; pp. 215–316; (b) Vibhute, A. M.; Komura, N.; Tanaka, H.-N.; Imamura, A.; Ando, H. Advanced Chemical Methods for Stereoselective Sialylation and Their Applications in Sialoglycan Syntheses. *Chem. Rec.* **2021**, *21*, 3194–3223.
3. Kanie, O.; Kiso, M.; Hasegawa, A. Glycosylation Using Methylthioglycosides of N-acetylneuraminic Acid and Dimethyl(methylthio)sulfonium Triflate. *J. Carbohydr. Chem.* **1988**, 7, 501–506.
4. Ando, H.; Koike, Y.; Ishida, H.; Kiso, M. Extending the Possibility of an *N*-Troc-Protected Sialic Acid Donor Toward Variant Sialo-glycoside Synthesis. *Tetrahedron Lett.* **2003**, *44*, 6883–6886.
5. Ando, H.; Koike, Y.; Koizumi, S.; Ishida, H.; Kiso, M. 1,5-Lactamized Sialyl Acceptors for Various Disialoside Syntheses: Novel Method for the Synthesis of Glycan Portions of Hp-s6 and HLG-2 Gangliosides. *Angew. Chem. Int. Ed.* **2005**, *44*, 6759–6763.
6. (a) Tamai, H.; Ando, H.; Tanaka, H.-N.; Hosoda-Yabe, R.; Yabe, T.; Ishida, H.; Kiso, M. The Total Synthesis of the Neurogenic Ganglioside LLG-3 Isolated From the Starfish *Linckia laevigata*. *Angew. Chem. Int. Ed.* **2011**, *50*, 2330–2333; (b) Goto, K.; Suzuki, T.; Tamai, H.; Ogawa, J.; Imamura, A.; Ando, H.; Ishida, H.; Kiso, M. Total Synthesis and Neuritogenic Activity Evaluation of Ganglioside PNG-2A From the Starfish *Protoreaster nodosus*. *Asian J. Org. Chem.* **2015**, *4*, 1160–1171.
7. Tanaka, H.; Ando, H.; Ishihara, H.; Koketsu, M. Sialylation Reactions With 5-*N*,7-*O*-Carbonyl-Protected Sialyl Donors: Unusual Stereoselectivity With Nitrile Solvent Assistance. *Carbohydr. Res.* **2008**, *343*, 1585–1593.
8. Crich, D.; Kancharla, P. K. Influence of Side Chain Conformation and Configuration on Glycosyl Donor Reactivity and Selectivity as Illustrated by Sialic Acid Donors Epimeric at the 7-Position. *J. Am. Chem. Soc.* **2013**, *135*, 18999–19007.

9. Crich, D.; Wu, B. Imposing the Trans/Gauche Conformation on a Sialic Acid Donor With a 5-*N*,7-*O*-Oxazinanone Group: Effect on Glycosylation Stereoselectivity. *Tetrahedron* **2008**, *64*, 2042–2047.
10. Tanase, M.; Imamura, A.; Ando, H.; Ishida, H.; Kiso, M. A 5-Ureido-Modified Sialyl Donor: A Tool for the Synthesis of α-Sialosides. *Org. Lett.* **2016**, *18*, 1454–1457.
11. Tanase, M.; Imamura, A.; Ando, H.; Kiso, M.; Ishida, H. Linear Synthesis of the Chol-1 Ganglioside Core Tetrasaccharide and Disialyl T Antigen Glycan Using a 5-Ureido-sialyl Donor. *Biosci. Biotech. Biochem.* **2017**, *81*, 2268–2278.
12. Komura, N.; Kato, K.; Udagawa, T.; Asano, S.; Tanaka, H.-N.; Imamura, A.; Ishida, H.; Kiso, M.; Ando, H. Constrained Sialic Acid Donors Enable Selective Synthesis of α-Glycosides. *Science* **2019**, *364*, 677–680.
13. Kraus, G. A.; Hon, Y.-S.; Thomas, P. J.; Laramay, S.; Liras, S.; Hanson, J. Organic Synthesis Using Bridgehead Carbocations and Bridgehead Enones. *Chem. Rev.* **1989**, *89*, 1591–1598.
14. (a) Smith, D. C.; Lord, J. M.; Roberts, L. M.; Johannes, L. Glycosphingolipids as Toxin Receptors. *Semin. Cell Dev. Biol.* **2004**, *15*, 397–408; (b) Lopez, P. H.; Schnaar, R. L. Gangliosides in Cell Recognition and Membrane Protein Regulation. *Curr. Opin. Struct. Biol.* **2009**, *19*, 549–557; (c) Fleming, F. E.; Bohm, R.; Dang, V. T.; Holloway, G.; Haselhorst, T.; Madge, P. D.; Deveryshetty, J.; Yu, X.; Blanchard, H.; von Itzstein, M.; Coulson, B. S. Relative Roles of GM1 Ganglioside, *N*-Acylneuraminic Acids, and α2β1 Integrin in Mediating Rotavirus Infection. *J. Virol.* **2014**, *88*, 4558–4571; (d) Blaum, B. S.; Frank, M.; Walker, R. C.; Neu, U.; Stehle, T. Complement Factor H and Simian Virus 40 Bind the GM1 Ganglioside in Distinct Conformations. *Glycobiology* **2016**, *26*, 532–539.
15. (a) Kawashima, N.; Yoon, S.; Itoh, J. K.; Nakayama, K. Tyrosine Kinase Activity of Epidermal Growth Factor Receptor is Regulated by GM3 Binding through Carbohydrate to Carbohydrate Interactions. *J. Biol. Chem.* **2009**, *284*, 6147–6155; (b) Fukuda, Y.; Fukui, T.; Hikichi, C.; Ishikawa, T.; Murate, K.; Adachi, T.; Imai, H.; Fukuhara, K.; Ueda, A.; Kaplan, A. P.; Mutoh, T. Neurotropin Promotes NGF Signaling Through Interaction of GM1 Ganglioside With Trk Neurotrophin Receptor in PC12 Cells. *Brain Res.* **2015**, *1596*, 13–21; (c) Jung, W. R.; Kim, H. G.; Shin, M. K.; Park, D. I.; Kim, K. L. The Effect of Ganglioside GQ1b on the NMDA Receptor Signaling Pathway in H19-7 Cells and Rat Hippocampus. *Neuroscience* **2010**, *165*, 159–167.
16. Kaneko, M.; Yamada, K.; Miyamoto, T.; Inagaki, M.; Higuchi, R. Neuritogenic Activity of Gangliosides From Echinoderms and Their Structure–Activity Relationship. *Chem. Pharm. Bull.* **2007**, *55*, 462–463.
17. Fujikawa, K.; Nakashima, S.; Konishi, M.; Fuse, T.; Komura, N.; Ando, T.; Ando, H.; Yuki, N.; Ishida, H.; Kiso, M. The First Total Synthesis of Ganglioside GalNAc-GD1a, a Target Molecule for Autoantibodies in Guillain-Barré Syndrome. *Chem. Eur. J.* **2011**, *17*, 5641–5651.
18. Konishi, M.; Imamura, A.; Fujikawa, K.; Ando, H.; Ishida, H.; Kiso, M. Extending the Glucosyl Ceramide Cassette Approach: Application in the Total Synthesis of Ganglioside GalNAc-GM1b. *Molecules* **2013**, *18*, 15153–15181.
19. Higuchi, R.; Inoue, S.; Inagaki, K.; Sakai, M.; Miyamoto, T.; Komori, T.; Inagaki, M.; Isobe, R. Biologically Active Glycosides From Asteroidea, 42. Isolation and Structure of a New Biologically Active Ganglioside Molecular Species From the Starfish *Asterina pectinifera*. *Chem. Pharm. Bull.* **2006**, *54*, 287–291.
20. Goto, K.; Sawa, M.; Tamai, H.; Imamura, A.; Ando, H.; Ishida, H.; Kiso, M. The Total Synthesis of Starfish Ganglioside GP3 Bearing a Unique Sialyl Glycan Architecture. *Chem. Eur. J.* **2016**, *22*, 8323–8331.

21. Sogabe, S.; Ando, H.; Koketsu, M.; Ishihara, H. A Novel De-*O*-Chloroacetylation Reagent: 1-Seleonocarbamoylpiperidine. *Tetrahedron Lett.* **2006**, *47*, 6603–6606.
22. (a) De Meo, C.; Demchenko, A. V.; Boons, G.-J. A Stereoselective Approach for the Synthesis of α-Sialosides. *J. Org. Chem.* **2001**, *66*, 5490–5497; (b) Tanaka, H.; Nishiura, Y.; Adachi, M.; Takahashi, T. Synthetic Study of α(2,8) Oligosialoside Using *N*-Troc Sialyl *N*-Phenyltrifluoroimidate. *Heterocycles* **2006**, *67*, 107–112; (c) Tanaka, H.; Nishiura, Y.; Takahashi, T. Stereoselective Synthesis of Oligo-α-(2,8)-Sialic Acids. *J. Am. Chem. Soc.* **2006**, *128*, 7124–7125; (d) Ando, H.; Shimizu, H.; Katano, Y.; Koizumi, S.; Ishida, H.; Kiso, M. Studies on the α-(1→4)- and α-(1→8)-Fucosylation of Sialic Acid for the Total Assembly of the Glycan Portions of Complex HPG-Series Gangliosides. *Carbohydr. Res.* **2006**, *341*, 1522–1532; (e) Tanaka, H.; Ando, H.; Ishida, H.; Kiso, M.; Ishihara, H.; Koketsu, M. Synthetic Study on α(2→8)-Linked Oligosialic Acid Employing 1,5-Lactamization as a Key Step. *Tetrahedron Lett.* **2009**, *50*, 4478–4481; (f) Hanashima, S.; Yamaguchi, Y.; Ito, Y.; Sato, K.-I. Synthesis of the Starfish Ganglioside AG2 Pentasaccharide. *Tetrahedron Lett.* **2009**, *50*, 6150–6153; (g) Hanashima, S.; Sato, K.; Naito, Y.; Takematsu, H.; Kozutsumi, Y.; Ito, Y.; Yamaguchi, Y. Synthesis and Binding Analysis of Unique AG2 Pentasaccharide to Human Siglec-2 Using NMR Techniques. *Bioorg. Med. Chem.* **2010**, *18*, 3720–3725; (h) Iwayama, Y.; Ando, H.; Tanaka, H.-N.; Ishida, H.; Kiso, M. Synthesis of the Glycan Moiety of Ganglioside HPG-7 With an Unusual Trimer of Sialic Acid as the Inner Sugar Residue. *Chem. Commun. (Cambridge)* **2011**, *47*, 9726–9728; (i) Xu, F.-F.; Wang, Y.; Xiong, D.-C.; Ye, X.-S. Stereoselective Synthesis of the Trisaccharide Moiety of Ganglioside HLG-2. *J. Org. Chem.* **2014**, *79*, 797–802.
23. Iwayama, Y.; Ando, H.; Ishida, H.; Kiso, M. A First Total Synthesis of Ganglioside HLG-2. *Chem. Eur. J.* **2009**, *15*, 4637–4648.
24. (a) Simons, K.; Ikonen, E. Functional Rafts in Cell Membranes. *Nature* **1997**, *387*, 569–572; (b) Simons, K.; Gerl, M. J. Revitalizing Membrane Rafts: New Tools and Insights. *Nat. Rev. Mol. Cell Biol.* **2010**, *11*, 688–699; (c) Sezgin, E.; Levental, I.; Mayor, S.; Eggeling, C. The Mystery of Membrane Organization: Composition, Regulation and Roles of Lipid Rafts. *Nat. Rev. Mol. Cell Biol.* **2017**, *18*, 361–374.
25. (a) Kusumi, A.; Suzuki, K. G. N. Toward Understanding the Dynamics of Membrane-Raft-Based Molecular Interactions. *Biochim. Biophys. Acta* **2005**, *1746*, 234–251; (b) Eggeling, C.; Ringemann, C.; Medda, R.; Schwarzmann, G.; Sandhoff, K.; Polyakova, S.; Belov, V. N.; Hein, B.; von Middendorff, C.; Schonle, A.; Hell, S. W. Direct Observation of the Nanoscale Dynamics of Membrane Lipids in a Living Cell. *Nature* **2009**, *457*, 1159–1162; (c) Sengupta, P.; Jovanovic-Talisman, T.; Skoko, D.; Renz, M.; Veatch, S. L.; Lippincott-Schwartz, J. Probing Protein Heterogeneity in the Plasma Membrane Using PALM and Pair Correlation Analysis. *Nat. Methods* **2011**, *8*, 969–975; (d) Suzuki, K. G. N.; Kasai, R. S.; Hirosawa, K. M.; Nemoto, Y. L.; Ishibashi, M.; Miwa, Y.; Fujiwara, T. K.; Kusumi, A. Transient GPI-Anchored Protein Homodimers are Units for Raft Organization and Function. *Nat. Chem. Biol.* **2012**, *8*, 774–783.
26. (a) Komura, N.; Suzuki, K. G. N.; Ando, H.; Konishi, M.; Koikeda, M.; Imamura, A.; Chadda, R.; Fujiwara, T. K.; Tsuboi, H.; Sheng, R.; Cho, W.; Furukawa, K.; Furukawa, K.; Yamauchi, Y.; Ishida, H.; Kusumi, A.; Kiso, M. Raft-Based Interactions of Gangliosides With a GPI-Anchored Receptor. *Nat. Chem. Biol.* **2016**, *12*, 402–410; (b) Komura, N.; Yamazaki, A.; Imamura, A.; Ishida, H.; Kiso, M.; Ando, H. Syntheses of Bifunctional Photoaffinity Ganglioside Probes for Studying Raft-Associated Interactions. *Trends Carbohydr. Res.* **2017**, *9*, 1–26; (c) Koikeda, M.; Komura, N.; Tanaka, H.-N.; Imamura, A.; Ishida, H.; Kiso, M.; Ando, H. Synthesis

of Ganglioside Analogs Containing Fluorescently Labeled GalNAc for Single-Molecule Imaging. *J. Carbohydr. Chem.* **2019**, *38*, 509–527.
27. Konishi, M.; Komura, N.; Hirose, Y.; Suganuma, Y.; Tanaka, H.-N.; Imamura, A.; Ishida, H.; Suzuki, K. G. N.; Ando, H. Development of Fluorescent Ganglioside GD3 and GQ1b Analogs for Elucidation of Raft-Associated Interactions. *J. Org. Chem.* **2020**, *85*, 15998–16013.
28. Asano, S.; Pal, R.; Tanaka, H.-N.; Imamura, A.; Ishida, H.; Suzuki, K. G. N.; Ando, H. Development of Fluorescently Labeled SSEA-3, SSEA-4, and Globo-H Glycosphingolipids for Elucidating Molecular Interactions in the Cell Membrane. *Int. J. Mol. Sci.* **2019**, *20*, 6187.

CHAPTER FOUR

Synthesis of homogeneous glycoproteins with diverse N-glycans☆

Kota Nomura†, Yanbo Liu†, and Yasuhiro Kajihara‡,*
Department of Chemistry Osaka University, Machikaneyama, Toyonaka, Osaka, Japan
*Corresponding author: e-mail address: kajihara@chem.sci.osaka-u.ac.jp

Contents

Abbreviations

Ac	acetyl
Acm	acetamidomethyl
Bn	benzyl
Boc	*tert*-butyloxycarbonyl
CCL-1	chemokine (C—C motif) ligand 1
CD	circular dichroism
CDAP	1-cyano-4-(dimethylamino)pyridinium tetrafluoroborate
CDF	cation diffusion facilitators
CSA	10-camphorsulfonic acid
Dbz	diaminobenzoic acid
DDC	diacyl disulfide coupling
DEPBT	3-(diethoxyphosphoryloxy)-1,2,3-benzotriazin-4(3*H*)-one
DIC	*N,N′*-diisopropylcarbodiimide
DIEA	*N,N*-diisopropylethylamine

† Equal contribution as the first author.
‡ Primary or first author.

☆Dedicated to the memory of Professor Hidetoshi Yamada.

Advances in Carbohydrate Chemistry and Biochemistry, Volume 81
ISSN 0065-2318
https://doi.org/10.1016/bs.accb.2022.09.004

DKP	diketopiperazine
DMF	*N,N*-dimethylformamide
DMS	dimethyl sulfide
E. coli	*Escherichia coli*
ELISA	enzyme-linked immuno sorbent assay
EPO	erythropoietin
ER	endoplasmic reticulum
Fmoc	9-fluorenylmethoxycarbonyl
GQC	glycoprotein quality control
HOBt	1,2,3-benzotriazol-1-ol monohydrate
HPLC	high-performance liquid chromatography
HSQC	heteronuclear single quantum coherence
ICOS	inducible T cell co-stimulator
IFN-β	interferon-β
IL	interleukin
LC–MS	liquid chromatography mass spectrometry
M9	mannose-9
MCP-3	monocyte chemotactic protein-3
Nbz	*N*-acyl-benzimidazoli-none
NCL	native chemical ligation
NIS	*N*-iodosuccinimide
NMR	nuclear magnetic resonance
NOESY	nuclear overhauser effect spectroscopy
ODS	octadecylsilyl
Pac	phenacyl
P/C-TM	post- and co-translational modification
PyBOP	benzotriazol-1-yloxytripyrrolidinophosphonium hexafluorophosphate
RP-HPLC	reversed-phase high-performance liquid chromatography
SEA	bis(2-mercaptoethyl)amine
SGI	specific and gradient of ^{15}N-isotope concentration
SPPS	solid-phase peptide synthesis
SUMO	small ubiquitin-like modifier
TCL	thioacid capture ligation
TFA	trifluoroacetic acid
TfOH	triflic acid
TIPS	triisopropylsilane
TOCSY	total correlation spectroscopy
UGGT	UDP-glucose:glycoprotein glucosyltransferase

1. Introduction

In the post-genomic era, post- and co-translational modifications (P/C-TMs) of proteins are known as more essential elements for protein function activation.[1] Recent research has revealed that these modifications regulate a wide range of biological processes, including several diseases such as cancer, Alzheimer's disease, and Parkinson's disease.[2–5] However, in recent years, P/C-TMs have been frequently focused on as not only essential basic biochemistry subjects but also as targets for drug discovery in the pharmaceutical industry.

Among these protein modifications, glycosylation is one of the most abundant modifications in eukaryotic cells. Glycoproteins are found on the cell surface and in cellular fluids, and these proteins are modified with serine/threonine-linked O-glycans or asparagine-linked N-glycans.[6] In case of N-glycans, the oligosaccharides are generally attached to the nitrogen of asparagine side chain that is present as a part of the consensus sequences; Asn-Xaa-Ser/Thr (Xaa: any amino acid except for Pro), whereby these glycans show considerable heterogeneity in their structure.[7] Structures of N-glycans are mainly classified as three types that include the complex-type, the high-mannose-type, and the hybrid-type (Fig. 1).

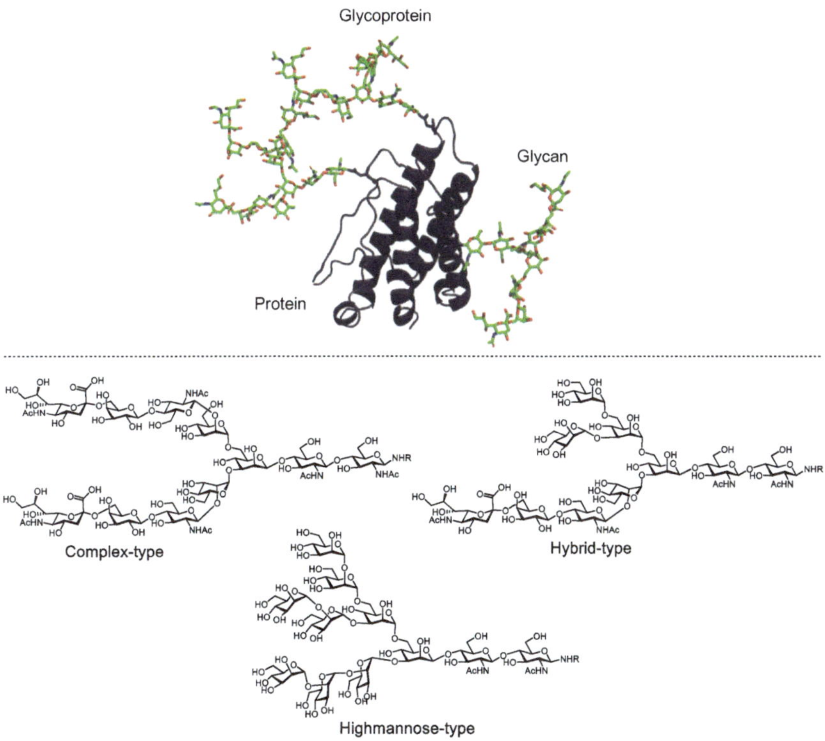

Fig. 1 Glycoprotein and glycan presented on the cell surface. Diversity of glycan structures presented on the glycoproteins. Complex-, hybrid-, and high-mannose-type glycans.

Such diverse glycosylation patterns increase the diversity of the corresponding glycoproteins' structure, and as a result, the functions of glycoproteins become more complex. However, the structural heterogeneity of glycosylation makes the functional analysis of individual glycosylation patterns very difficult. Unfortunately, preparation of a glycoprotein by cell expression methods affords heterogeneous products, and current purification techniques make it extremely difficult to isolate sufficient amounts of proteins with homogeneous glycan structures. The synthesis of glycoproteins with uniform glycan structures is essential for functional analysis of glycoproteins in biochemical research. For that purpose, chemical methods to synthesize glycoproteins with chemically uniform glycan structures have been developed.

The total chemical synthesis of proteins requires solid-phase peptide synthesis to obtain several peptides, and subsequent ligations to assemble these peptides sequentially and afford the desired full-length protein. Past decades have witnessed the development of protein chemical synthesis based on advanced solid-phase peptide synthetic protocols and newly established chemical ligation methods. Among these novel ligation methods, native chemical ligation (NCL), which enables the chemo-selective coupling between unprotected peptides, was established and widely used. The development of NCL by Dawson and coworkers greatly propelled the chemical synthesis of (glycol)proteins.[8–10] The reaction is performed between two fragments: a peptide thioester and an N-terminal cysteinyl peptide fragment, to afford a natural amide bond (Scheme 1). In principle, NCL occurs through a two-step process including capture and intramolecular S—N acyl shift. The capture step relies on a unique chemoselective reaction, such as a thiol–thioester exchange, that brings two unprotected peptides together. This step induces a subsequent intramolecular S—N acyl rearrangement reaction, leading to the regioselective amide bond formation. The unique mechanism of NCL enables couplings between unprotected peptides that have potentially reactive functional groups such as hydroxy, carboxy, and amino groups on their amino acid residues. However, using original

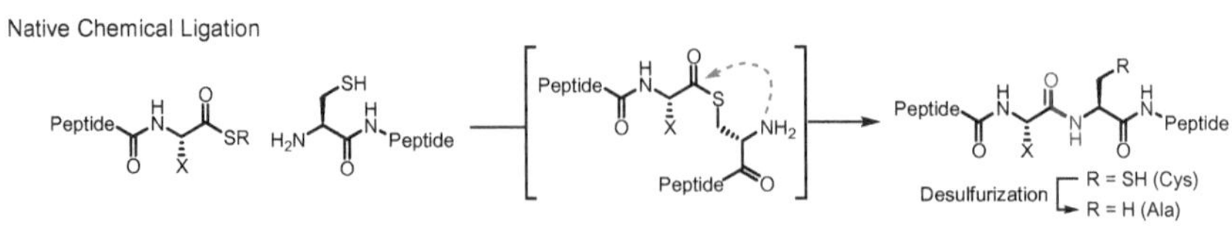

Scheme 1 Mechanism of native chemical ligation.

NCL, the ligated peptides have a Cys residue beside the ligated site. The Cys residue, which remains at the ligation site following the reaction, can be manipulated by hydrogenation or radical-based desulfurization to convert Cys residues to Ala, which were reported by the Dawson and then the Danishefsky groups.[11,12] This method has been widely adopted, for example in the total chemical synthesis of several complex proteins and glycoproteins. Moreover, several groups developed the synthesis of mercapto amino acids having β, γ, and δ-mercapto groups.[13,14] After ligation, the thiol group is selectively removed by radical desulfurization to restore the respective native amino acid residues in the sequence. By applying synthetic β-mercapto amino acid derivatives to the Cys residue of NCL, these workers demonstrated that desulfurization of this residue within the ligation products was possible to convert to any desired amino acids. Currently many groups have developed the synthetic method of mercapto amino acid derivatives.[15] Almost all the mercapto amino acids were successfully synthesized in the sophisticated reactions. Theoretically any amino acid is usable for the ligation reactions. NCL strategy has been further expanded by the application of the ligation–desulfurization strategy and applied for many syntheses of glycopeptides.[7,16]

In this review, we highlight our recent advances in preparation of homogeneous glycoproteins, including chemical synthesis of glycoprotein based on the synthesis of diverse homogeneous N-glycans, as well as semi-synthesis of glycoprotein based on newly developed methods for expressed peptide thioesterification.

2. Synthesis of homogeneous N-glycans and N-glycopeptides

The chemical synthesis of non-glycosylated proteins was greatly propelled by development of native chemical ligation (NCL); however, the synthesis of glycoproteins was still challenging in that it requires an efficient synthesis of homogeneous N-glycans and glycopeptides. The conventional total chemical synthesis of N-glycans is usually tedious and of low-efficiency due to structural diversity and the presence of some difficult-to-synthesize glycosidic bonds. In addition, the N-glycan is often not compatible with solid-phase peptide synthesis (SPPS) for the synthesis of glycopeptides on a resin, which requires strong acidic conditions for global deprotection.

In the past twenty years, the synthesis of homogenous glycoproteins is greatly facilitated by semi-synthesis of various N-glycans using sialylglycopeptides obtained from egg yolk. In 1997, the Seko group reported the extraction of sialylglycopeptide and sialylglycan from egg yolk, which paved the way for the efficient and large-scale synthesis of homogeneous N-glycans for glycoprotein synthesis.[17] However, at that time, the N-glycan structures available for synthesis were limited to only disialylglycans.

In order to expand the scope to obtain structurally diverse homogeneous N-glycans based on sialylglycan isolated from egg yolk, the Kajihara group reported a semi-synthetic method utilizing enzymatic digestion to obtain 24 types of N-glycan structures.[18,19] To achieve the branch-selective enzymatic digestion, selective acidic hydrolysis of only one sialic acid residue on disialo N-glycan **1** was carried out. The two types of monosialylglycans **2** and **3** that were obtained, along with disialylglycan **1** and asialoglycan **4** in the mixture, were purified by ODS-HPLC column chromatography after protection with the hydrophobic Fmoc group. With two kinds of monosialylglycans **2** and **3** in hand, the subsequent exo-glycosidase digestion gave diverse N-glycans after sequential digestion of the galactose, glucosamine and mannose residues with branch-specificity (structures **5–13**, Scheme 2). Finally, the author reported synthesis of 24 types of homogeneous N-glycans based on this approach on large scale. These homogeneous N-glycans can be used for SPPS of glycopeptides. By synthesizing structurally diverse homogeneous N-glycans, this research laid the foundation for the investigation of the individual structure–function relationships of N-glycans.

Triantennary N-glycans might affect the biological activities or physiological properties of glycoproteins compared to common biantennary N-glycans. However, the total chemical synthesis of homogeneous complex-type triantennary glycans was difficult due to multiple steps and regioselectivity. In 2016, the Kajihara group reported the efficient semi-synthesis of triantennary N-glycans based on asialoglycans isolated and prepared from egg yolk.[20] The main challenge the author needs to solve is how to attach the lactosamine **18** regio-selectively to one of 24 hydroxy groups on biantennary N-glycan **14**. To meet this challenge, a selective protection/deprotection protocol was established as described in Scheme 3.

First, the 4,6-diol of two galactose and two mannoses were selectively protected with 4,6-*O*-benzylidene groups, followed with Pac protection of asparagine and acetyl protection of the remaining hydroxy groups on the N-glycan. Next, selective deprotection of the 4,6-*O*-benzylidene group

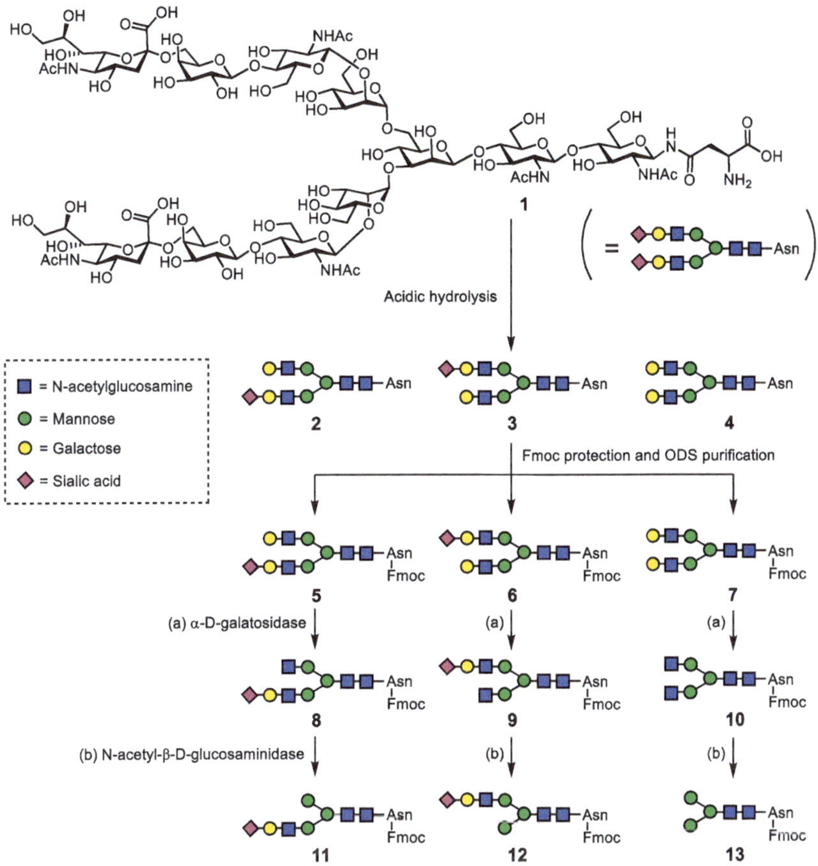

Scheme 2 Semi-synthesis of various complex-type N-glycans for the solid-phase synthesis of glycopeptides.

on the mannose residue, which will produce the future glycosylation site(s) for a triantennary N-glycan, was examined. In this key reaction, the authors found that the 4,6-*O*-benzylidene group of mannose is more prone to deprotection than that of galactose. Based on this reaction, formation of two kinds of biantennary acceptors **16** and **17** by selective deprotection of only one benzylidene group on mannose, was observed by LC–MS; the product was purified by RP-HPLC. (Scheme 3) The structure was characterized by 2D NMR spectroscopy, which indicated the selective deprotection of one benzylidene group on the mannose unit. Finally, glycosylation of two acceptors with lactosamine donor **18** produced, after deprotection, the desired triantennary N-glycans **21** and **22**. This

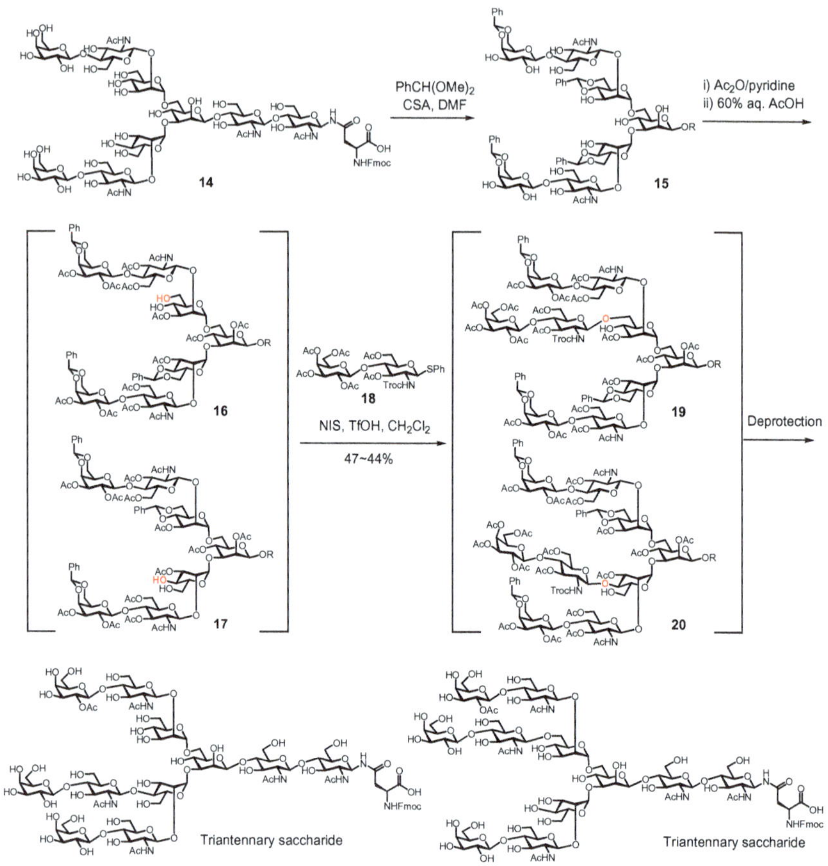

Scheme 3 Semi-synthesis of complex type triantennary glycans.

research paved the way for the synthesis of homogeneous glycoproteins containing triantennary N-glycans for the investigation of structure–function relationships.[21]

Although various homogeneous N-glycans were prepared on large scale, their application in SPPS was still troublesome for two reasons: (1) the acid-labile sialyl linkage and (2) the lack of proper coupling conditions of glycosyl asparagine with a peptide on a solid support. In order to incorporate semi-synthesized disialylglycan into SPPS efficiently, the Kajihara group reported a method for SPPS-compatible disialylglycopeptide synthesis by utilizing benzyl protection of the sialic acid and optimized coupling conditions.[22] Firstly, the coupling reagents for coupling of glycosylasparagine with **23** on a resin were screened to avoid undesired aspartimide formation.

It was found that 3-(diethoxyphosphoryloxy)-1,2,3-benzotriazin-4(3*H*)-one (DEPBT) gave the best result. Next, the coupling reaction of subsequent amino acids with resin-bound glycopeptide **24** was also optimized to avoid esterification of the free-OH group. When the DIC/HOBt reagent was employed for activation under low concentration, the esterification was substantially suppressed compared to high concentration conditions. Based on these optimized coupling conditions, a glycopeptide containing sialylglycan was synthesized by SPPS and NCL (Scheme 4). This research propelled the development of sialylglycoprotein synthesis by achieving the efficient synthesis of glycopeptides via SPPS.

Although the protocol for solid-phase synthesis of glycopeptides was established, the chemical synthesis of glycopeptides in the thioester form was still challenging. The general method toward peptide thioester synthesis relied on Boc-SPPS that required strongly acidic conditions (HF) for global deprotection, which might lead to the cleavage of the acid-labile sialyl linkage. In 2008, the Kajihara group reported two methods for synthesizing glycopeptide thioesters: (1) a minimal protection strategy based on Boc-SPPS directly, and (2) a coupling strategy using peptides derived from Fmoc-SPPS.[23] (Scheme 5)

In the first method (Scheme 5, left), in order to avoid strongly acidic conditions (HF) for side-chain deprotection during Boc-SPPS, the optimized peptide-coupling conditions that avoid esterification is utilized. As a result, the protection on amino acid side chains, especially on Ser and Thr, is avoided. After elongation on the resin, the glycopeptide **27** was cleaved by thioester exchange to afford the desired thioester **28** without any strongly acidic treatment. The purity of the glycopeptide thioester **28** that was obtained was determined by ^{1}H NMR spectroscopic measurements. In the second method (Scheme 5, right), the glycopeptide **29** was prepared by Fmoc-SPPS with protection on the side chains and a free C-terminal carboxylic acid. The C-terminus was converted to thioester by a coupling reaction with the corresponding thiol. Optimized conditions (PyBOP/DIEA, −20 °C) were utilized to avoid epimerization. Finally, the protecting groups on side chains were removed by treatment with a TFA cocktail to afford the desired glycopeptide thioester **30**. With the glycopeptide thioester (sequence:1–10, glycosylation: Asn6) in hand, a full-length glycoprotein MCP-3 is chemically synthesized by sequential NCL.[23] In this research, synthesis of homogeneous glycoproteins having N-glycans was achieved based on two new methods for obtaining glycopeptide thioesters. However, more efficient and general synthetic methods for glycopeptide thioesters were still in need.

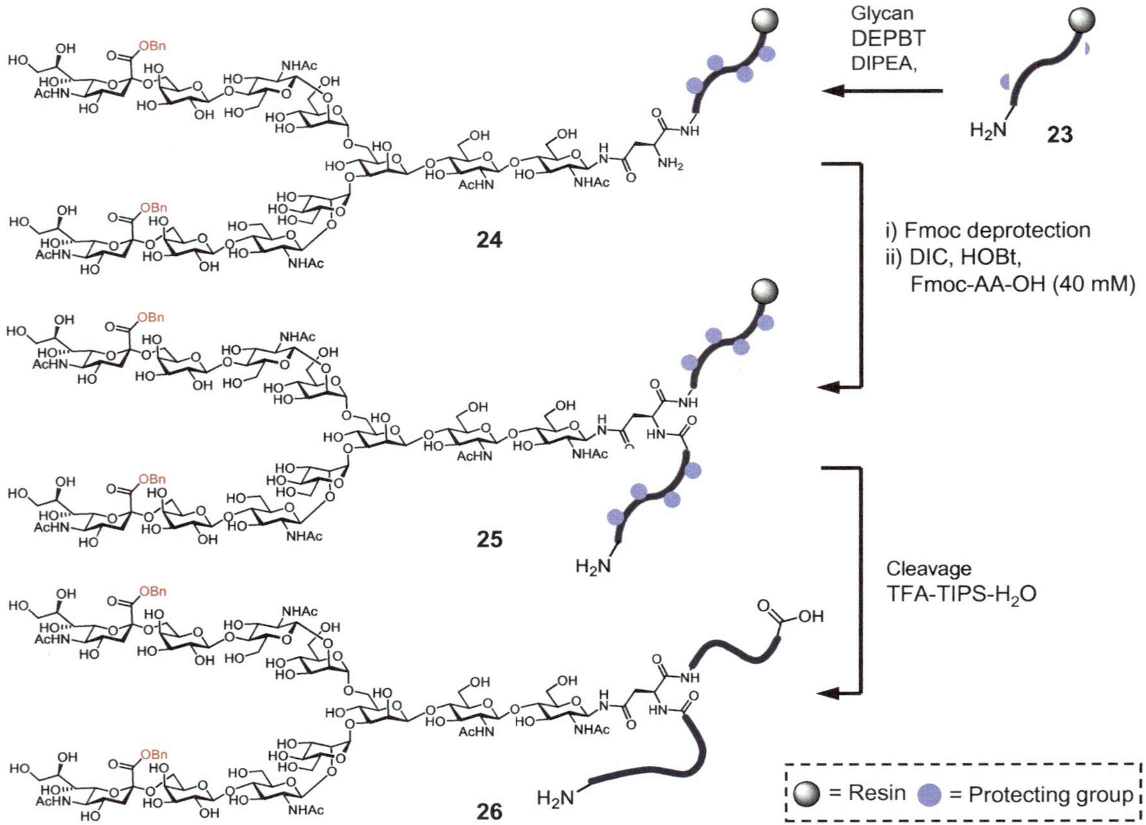

Scheme 4 Protocol to incorporate disialo complex-type N-glycans into Fmoc-SPPS for the synthesis of glycopeptides.

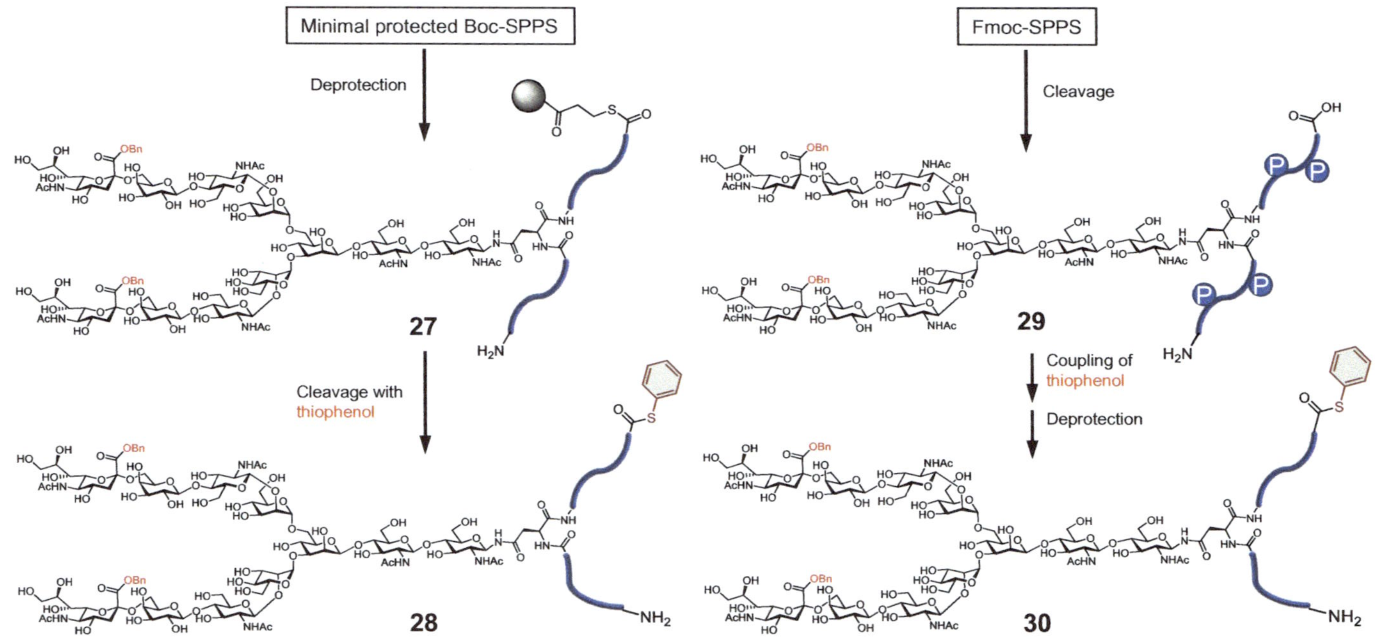

Scheme 5 Synthesis of glycopeptide thioesters based on Boc-SPPS and Fmoc-SPPS.

To develop a more efficient method for the synthesis of glycopeptide thioesters, Maki and coworkers reported a liquid-phase glycopeptide synthesis.[20,21] In this method, instead of coupling on a resin, the N-glycan was incorporated into peptide chain by a two-step sequential coupling in the liquid phase. For that purpose, the C-terminal partially protected peptide **32** prepared by SPPS was cleaved from the resin before coupling with glycosylasparagine **31**. The coupling reaction was carried out in the liquid phase to afford the desired glycopeptide **33**. After Fmoc deprotection, the glycopeptide **34** was further coupled in the liquid phase with N-terminal peptide **35** derived from SPPS. Since both coupling reactions were carried out in liquid phase, the synthetic efficiency was substantially improved compared with solid-phase coupling, which requires an excessive amount of glycosylasparagine. The glycopeptide **36** thus obtained was finally deprotected by a TFA cocktail to afford the desired glycopeptide **37** with a triantennary glycan (Scheme 6). The authors further succeeded in synthesizing erythropoietin with a complex-type triantennary sialylglycan.[21] The sialic

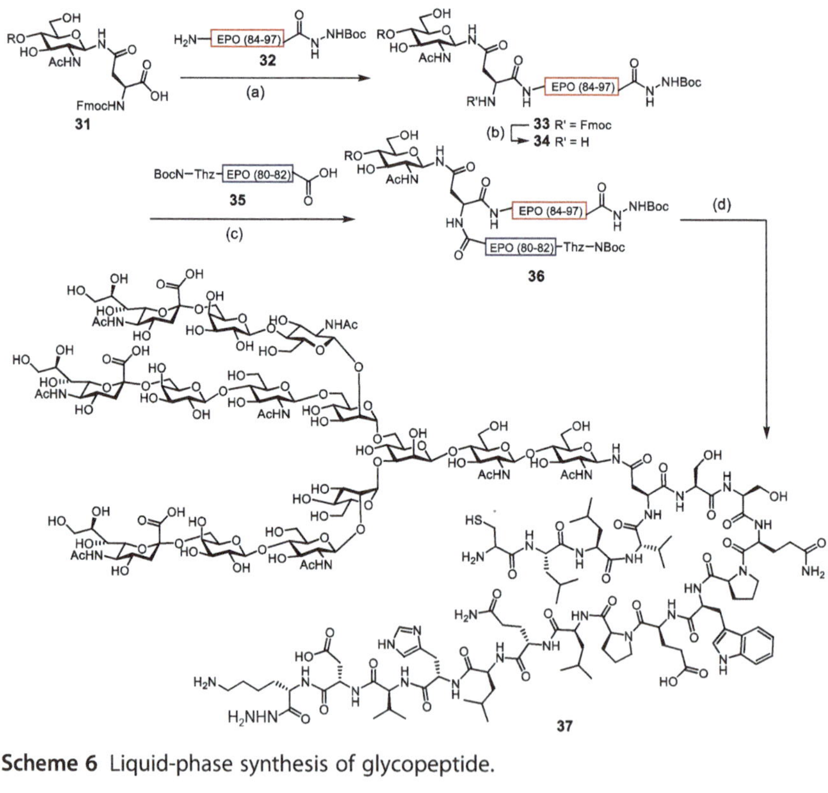

Scheme 6 Liquid-phase synthesis of glycopeptide.

acid was introduced by α-2,6-sialyltransferase. Since this reaction is carried out in the liquid phase, the condensation rate of peptide and glycosylasparagine was higher than that of SPPS. And since the condensation could be carried out at a low temperature, the rate of epimerization was reduced.

3. Total chemical synthesis of homogeneous N-glycoproteins

The chemical synthesis of glycoproteins is greatly facilitated by the development of native chemical ligation (NCL) along with methods for obtaining homogeneous, complex-type N-glycans from egg yolk. Based on these methods, in 2008, the Kajihara and Dawson group reported the chemical synthesis of glycoprotein monocyte chemotactic protein-3 (MCP-3), a CC-chemokine that consists of 76-amino acids and a complex-type N-glycan (Scheme 7).[23] The two steps of sequential NCLs using three unprotected peptide segments **38**, **39** and **40** afforded native glycoprotein MCP-3 **43**. The challenging synthesis of the glycopeptide thioester **38** was achieved by using either Fmoc SPPS protocol or Boc SPPS protocolas we mentioned in earlier (see Scheme 7). Using these strategies, they successfully ligated chemically synthesized peptide segments by NCL and examined oxidative folding conditions for the linear glycopeptides to afford a glycoprotein having a homogeneous complex-type sialylglycan and two disulfide bonds. Finally, they analyzed the positions of the disulfide bonds using enzymatic digestion. The CD and MS spectra, along with a simple ELISA assay, all indicated a completed synthesis of the desired glycosylated MCP-3.

Human interferon-β is a naturally N-glycosylated cytokine with immunomodulating bioactivity. The N-glycosylation of IFN-β is reported to affect the pharmacokinetics of this glycoprotein. In 2012, the Sakamoto and Kajihara group reported the chemical synthesis of the homogeneous glycoprotein interferon-β (IFN-β).[24] The synthetic sequence was divided into three segments (sequence: 1–67, 68–88, and 89–166) that could be assembled by native chemical ligation. The glycopeptide **44** (sequence: 68–88) containing a complex-type N-glycan at Asn80 was prepared by Fmoc-SPPS as the thioester form. After NCL to afford full-length IFN-β bearing either disialylglycan or asialylglycan, the folding via dialysis under redox conditions gave correctly folded IFN-β **48** for biological analysis. (Scheme 8). The biological activities of the two glycoforms were compared with commercially available human IFN-β with a heterogeneous N-glycan

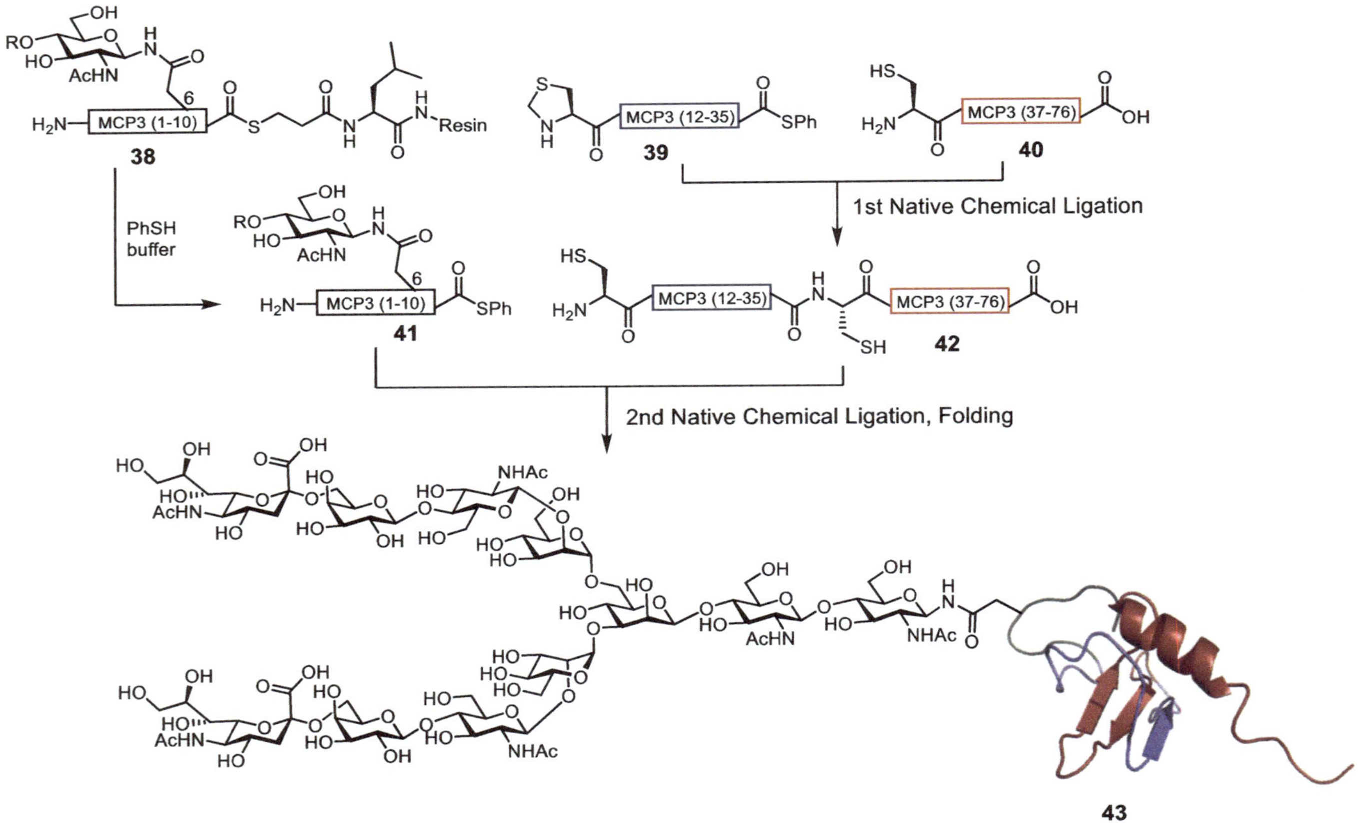

Scheme 7 Synthesis of MCP-3 having biantennary sialyloligosaccharides.

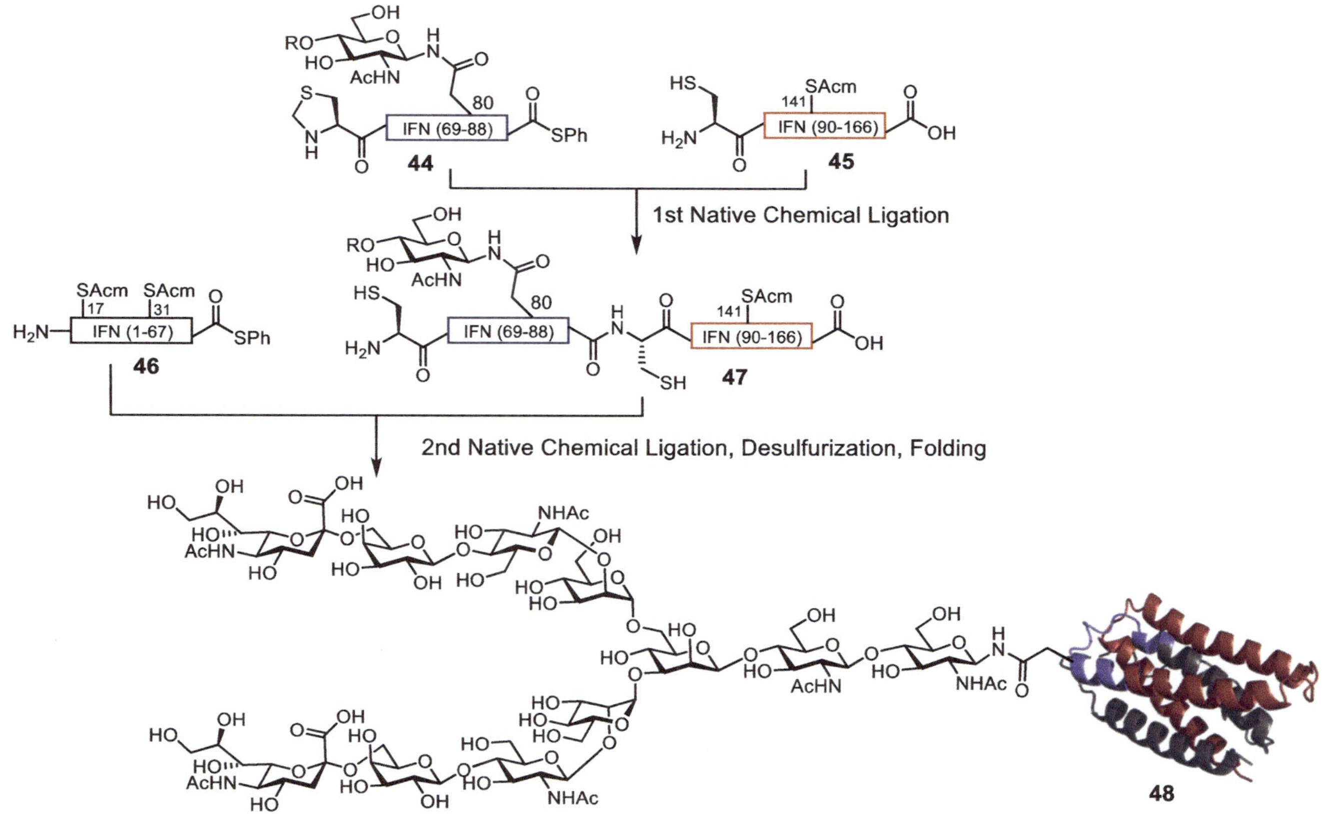

Scheme 8 Synthesis of interferon-β.

structure. Both antitumor activity assays in vitro and in vivo displayed similar bioactivity between synthetic and commercially available IFN-β. This was the first report to demonstrate that a synthetic glycoprotein having homogeneous N-glycan structure showed potent bioactivity in vivo. This result indicated that the synthetic homogeneous glycoprotein can be applied as a powerful tool for biological analysis to investigate structure–function relationships of N-glycans.

Erythropoietin (EPO) is a hormone that plays important roles in erythropoiesis. The sequence of EPO consists of 166 residues, including three N-glycosylation sites at Asn24, 38 and 83. Although EPO performs important biofunctions as a glycoprotein, the function of its N-glycans is still not fully understood. The synthesis of homogeneous N-glycosylated EPO for elucidation of its N-glycan function has been studied by chemists. In 2012, Murakami and coworkers achieved the synthesis of folded erythropoietin bearing a complex-type sialylglycan at Asn83 (Scheme 9).[25] The traditional Boc-SPPS method was challenging for synthesizing the disialo-glycopeptide thioester due to the acid-labile feature of the sialic acid residue. To solve this problem, a synthetic strategy was developed based on phenacyl (Pac)-protection of sialic acid on the N-glycan. The N-glycosylated asparagine derived from egg yolk was treated with phenacyl bromide for protection before being incorporated into SPPS. After coupling on resin, the protected disialo N-glycan kept intact under acidic conditions for both t-Boc deprotection (TFA) and global deprotection (TFA/TfOH/DMS/*m*-cresol). Finally, to test the practicability of this method, glycoprotein EPO having a disialo N-glycan at Asn83 was chemically synthesized. The glycopeptide thioester **57** (sequence:79–97) bearing a Pac-protected disialo N-glycan was smoothly synthesized by t-Boc-SPPS for full-length EPO synthesis. The four native cysteines were protected by the Acm protecting group, and alternatively five alanine residues were selected as ligation sites for the six fragments. Four segments **56**, **57**, **58** and **59** were ligated sequentially by NCL. The glycopeptide (sequence: 50–166) was subjected to desulfurization of the three free thiol groups to afford glycopeptide **60**. A final ligation and deprotection of the Acm protecting group of the four native cysteines gave the linear EPO. The oxidative folding protocols that followed afforded an EPO **61** with a disialylglycan. CD spectra and MS spectra suggested the correct folding.

Then in 2016, Murakami and coworkers succeeded in the synthesis of the glycoproteins of EPO in which the N-glycosylation numbers and positions were varied from one to three.[26] In this synthesis, they installed

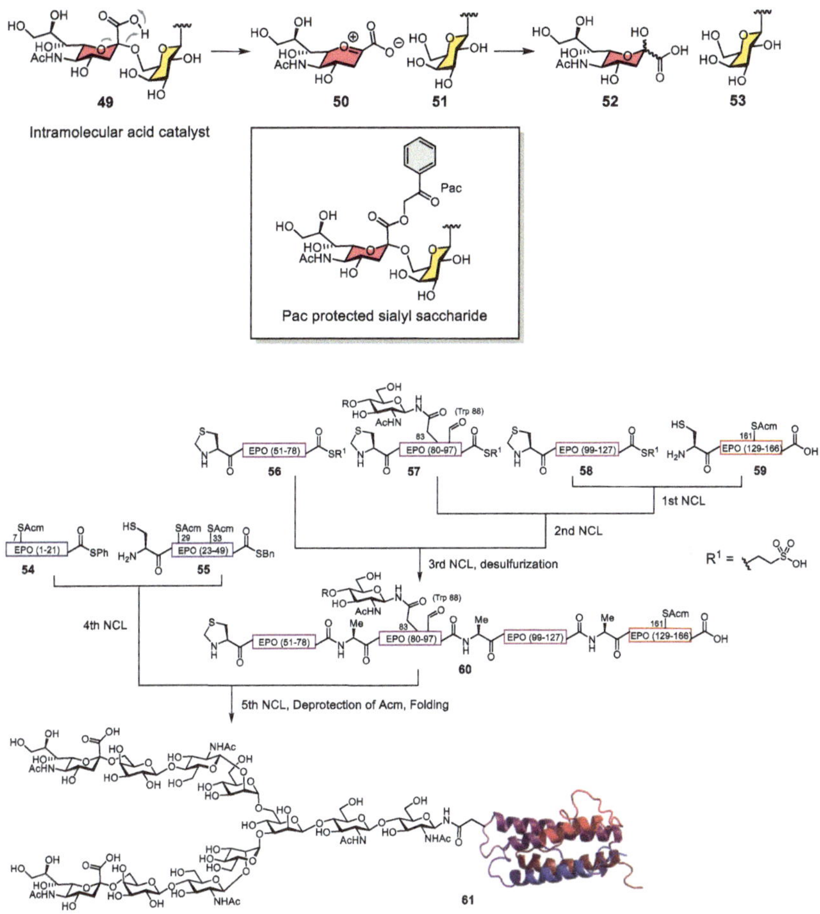

Scheme 9 Synthesis of erythropoietin with a biantennary sialylglycan.

complex-type biantennary sialylglycans at Asn24, Asn38, and Asn83, and as a result, they synthesized five EPO glycoforms varying in glycosylation positions and in the numbers of human-type biantennary sialylglycans (Scheme 10). They synthesized glycopeptides **62**, **63** and **65** having complex-type, Pac-protected sialylglycan by a standard Boc-SPPS protocol for the synthesis of the glycopeptide thioester building block. Then six segments were sequentially coupled by NCL and followed by desulfurization. Finally, deprotection of the Acm protecting groups afforded the linear, full-length, glycosyl polypeptide. After folding under stepwise dialysis conditions, five types of the folded glycosyl EPO glycoforms

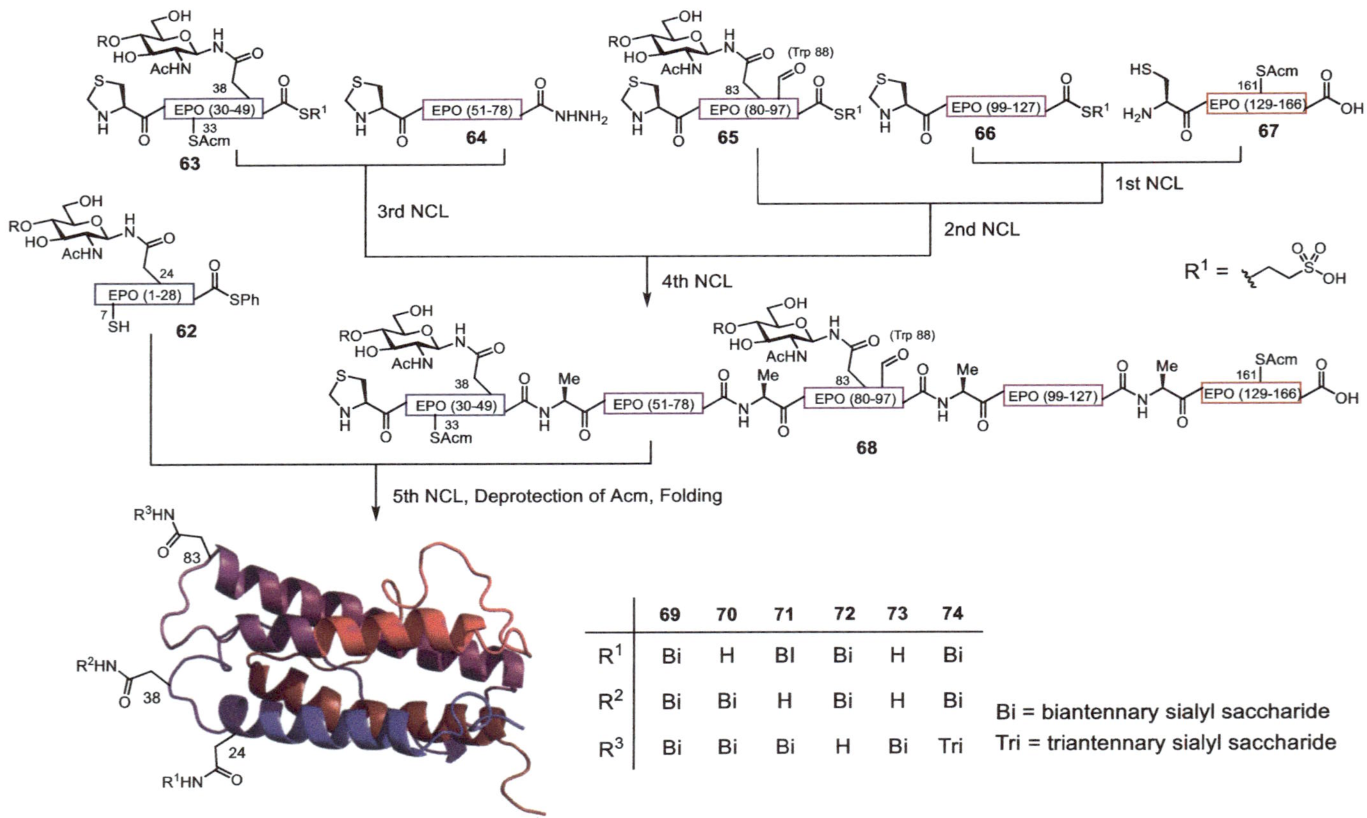

	69	70	71	72	73	74
R^1	Bi	H	BI	Bi	H	Bi
R^2	Bi	Bi	H	Bi	H	Bi
R^3	Bi	Bi	Bi	H	Bi	Tri

Bi = biantennary sialyl saccharide
Tri = triantennary sialyl saccharide

Scheme 10 Synthesis of five glycoforms of EPO varying in glycosylation positions and the numbers of human-type biantennary sialylglycan.

69–73 were obtained. Based on an in vivo erythropoiesis assay in mice, all types of the EPOs displayed biological activities. Especially the EPOs having three glycan chains, EPO **69**, exhibited the highest in vivo activity. Furthermore, they discussed structure–activity relationships. Consequently, this example became the first work to give an insight into the relationship between glycosylation patterns and hematopoietic activity. In addition to this synthesis, based on the same assembly strategy, Maki and Kajihara achieved the installation of a complex-type triantennary sialylglycan at Asn83, and as a result they synthesized alternative EPO glycoforms with a triantennary sialylglycan such as **74**.[21]

CCL-1 is a member of the chemokine family and is naturally N-glycosylated. CCL-1 plays an important function in chemotaxis and apoptosis; however, the function of its N-glycan is poorly understood. In 2014, the Okamoto and Kent group reported the chemical synthesis of homogeneous N-glycosylated CCL1 and Ser-CCL1 for investigation of their biological activities.[27] In order to minimize the procedure for glycopeptide synthesis, the sequence of CCL1 was divided into three segments: (1) peptide thioester **75** (sequence:1–25), (2) short glycopeptide thioester **76** (26–33, Asn29 N-glycosylated), and (3) peptide **77** (sequence:34–73). The complex-type asialo N-glycan-Asn derived from egg yolk was incorporated into Fmoc SPPS for glycopeptide thioester synthesis. Although traditional Fmoc SPPS could not produce the glycopeptide thioester, the use of diaminobenzoic acid (Dbz) as a linker solved this problem and gave glycopeptide-Nbz **76** for NCL. The convergent NCL using three segments smoothly afforded full-length glycosylated CCL-1 **79** (Scheme 11). Following the same protocol, non-glycosylated CCL1 and glycosylated and non-glycosylated Ser-CCL1 were also efficiently synthesized. With four synthetic (glyco)proteins in hand, their biological activities were tested by a chemotaxis assay using a CCR8-transfected cell line for the elucidation of N-glycan function. According to the result, intriguingly, N-glycosylated CCL1 showed decreased chemotaxis activity compared with non-glycosylated CCL1, which indicates that N-glycosylation might affect other aspects of CCL1 biofunctions.

For elucidating the effect of N-glycans on glycoproteins, Hien and coworkers synthesized fractalkine (CDF) having a complex-type and high-mannose-type glycans (Scheme 12).[28] Furthermore, they conducted structural analyses of the glycoproteins using stable isotope labeling with labeling of Specific position and a Gradient of ^{15}N-Isotope concentrations (SGI). CDF consists of 76 amino acids and has N-glycan at Asn9. According

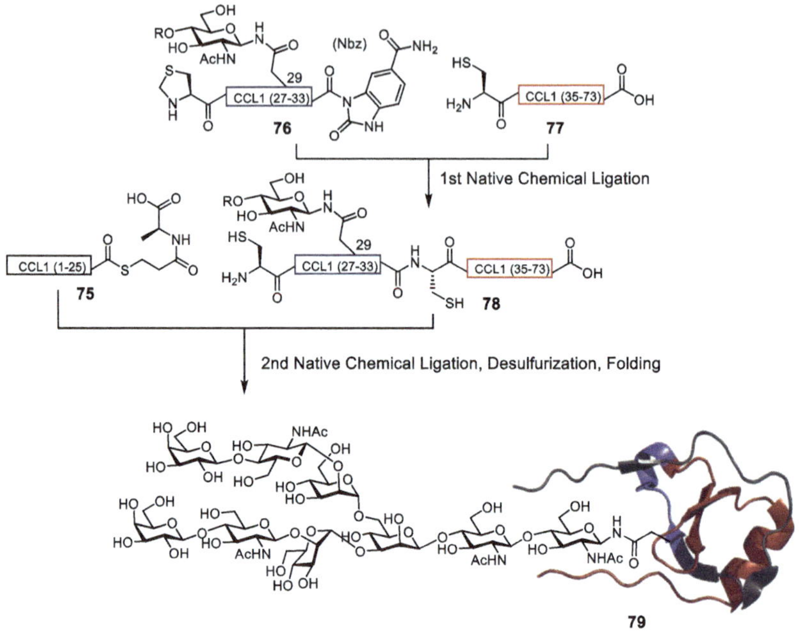

Scheme 11 Synthesis of CCL1 having a biantennary asialoglycan.

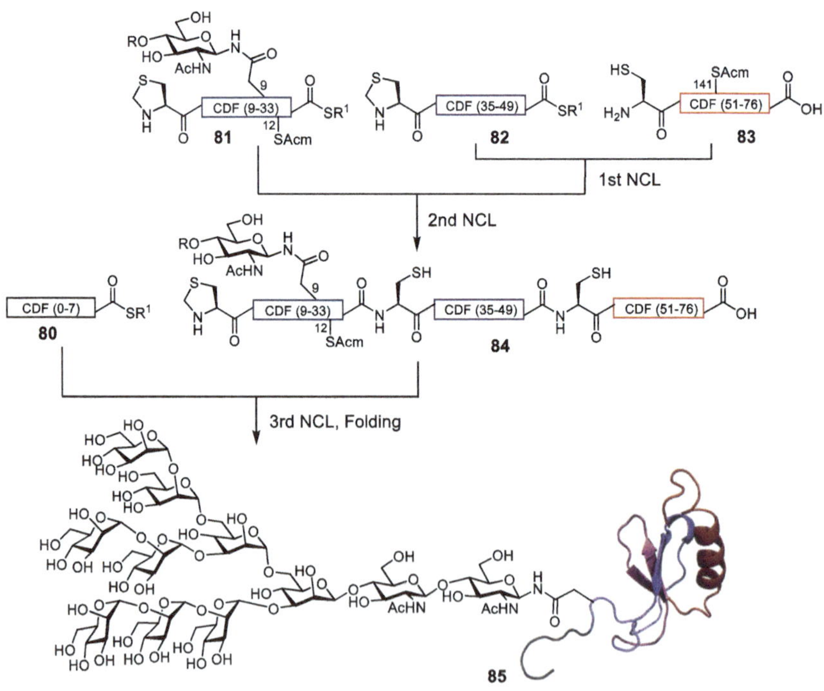

Scheme 12 Synthesis of CDF and structural analysis by using specific labeling and gradients of ^{15}N-isotope concentration

to their synthetic scheme, these workers synthesized glycopeptide thioester **81** having a complex-type nonasaccharide and a high-mannose-type saccharide by either standard Fmoc or Boc protocols, respectively. Then four segments were sequential coupled by NCL and then deprotection were performed. Finally, deprotection of the Acm protecting groups afforded the full-length glycopeptide. After the oxidative folding process two types of folded glycosyl CDFs **85** were produced in which thirteen ^{15}N-labeled amino acids were inserted at specific positions. They also synthesized non-glycosyl CDF in the same manner. CD spectra and enzymatic disulfide bond mapping confirmed that both glycosylated CDFs **85** and non-glycosylated CDF were correctly folded. The SGI labeling protocol enabled them to assign specific amino acids to the ^{1}H-^{15}N HSQC spectra and investigate the dynamics of local conformational properties of the three synthetic CDFs. According to the HSQC studies combined with TOCSY and NOESY methods, the glycosylations did not affect conformational changes of CDF at room temperature. However, T1 values indicated that the glycan motif influenced the dynamics at the local conformation. As a result, temperature-varied circular dichroism CD spectra and T1 values indicated that N-glycans might inhibit protein fluctuation and stabilize protein structure.

In the previous chemical synthesis, the N-glycan part was usually incorporated by SPPS, which causes waste of precious N-glycans during the coupling steps. To solve this problem, a new synthetic strategy without using SPPS to obtain glycopeptide might be a powerful tool toward glycoprotein synthesis. In 2021, the Nomura and Kajihara group reported a highly efficient synthetic strategy by the chemical insertion of N-glycosylated asparagine based on a bifunctional thioacid derivative.[29] In this synthetic strategy, instead of incorporating N-glycan by SPPS, the N-glycan is inserted utilizing N-glycosylated asparagine thioacid **86** as a ligation junction. The full-length glycoprotein could be efficiently prepared after only two coupling reactions: (1) In the first step diacyl disulfide coupling (DDC), the N-terminal peptide thioacid **87** was coupled with N-glycosylated asparagine thioacid **86** under oxidative condition to afford the desired glycopeptide **88** as a thioacid form (Scheme 13). The reaction was carried out in a chemoselective and epimerization-free manner without undesired coupling byproducts. In the second step, the coupling product **88** was ligated with C-terminal peptide **89** via thioacid capture ligation (TCL)[30,31] through a nitro pyridyl sulfide (Npys) modified N-terminal Cys (Scheme 13). Based on such N-glycan insertion strategy, the authors

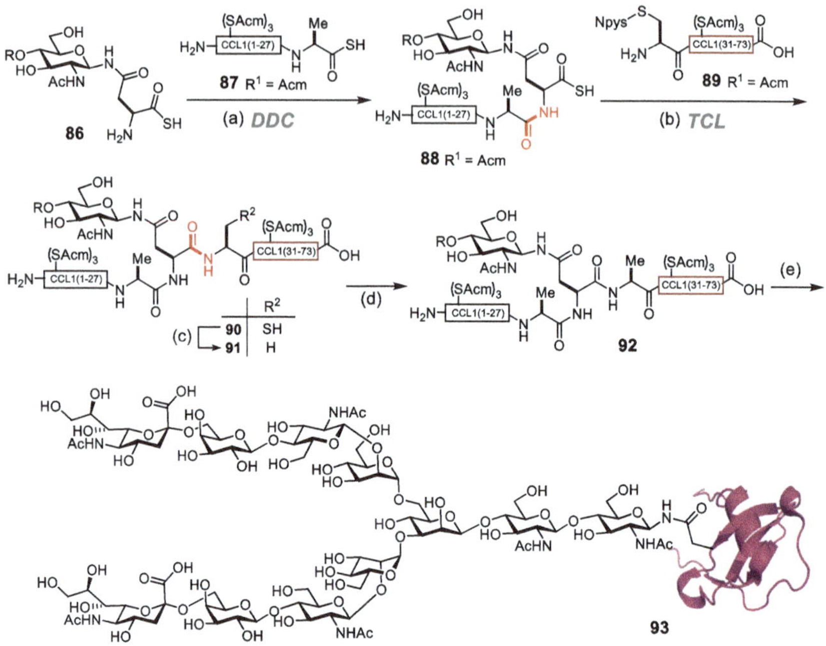

Scheme 13 Synthesis of CCL-1 using a bifunctional thioacid-mediated strategy.

efficiently synthesized full-length glycoprotein CCL-1 **93** and interleukin-3 (IL-3) in only a few chemical conversion steps. This novel thioacid-mediated strategy provides a robust option for the convergent synthesis of glycoproteins without the need for preparing glycopeptides by SPPS.

With the development of ligation methods and metal-free desulfurization, total chemical synthesis became a powerful tool for the synthesis of homogeneous glycoproteins, and it has already succeeded on several targets including EPO, IFN-β, CCL-1, MCP-3, IL-3 and CDF. The main challenge that remains is still how to synthesize the glycopeptides efficiently. At present, the SPPS-based method, the Endo M-catalyzed method, the aspartylation method, and the liquid-phase synthesis have been developed and utilized; however, more efficient methods or a general synthetic strategy is still in need.

By chemical synthesis of homogeneous glycoproteins, the glycan function, especially individual structure–function relationships, could be better understood. Homogeneous glycoproteins bearing distinct glycan structures and positions, that is, distinct glycoforms, could be efficiently prepared and used for the investigation of glycan functions. Synthetic homogeneous glycoproteins can also be compared with commercially available heterogeneous glycoproteins to evaluate their bioactivities.

4. Investigation of N-glycosylation functions utilizing synthetic glycoproteins

N-Glycosylation is known to play a pivotal role in the glycoprotein quality control system, especially by recognition with UDP-glucose glycoprotein glucosyltransferase (UGGT) as a folding sensor. UGGT can recognize misfolded glycoproteins with a high-mannose glycan and then add a glucose unit as a tag for the misfolded glycoprotein in the endoplasmic reticulum (ER). However, how UGGT recognizes and interacts with a misfolded glycoprotein is poorly understood due to lack of homogeneous misfolded glycoproteins for investigation. In 2012, the Kajihara group reported the chemically synthesized homogeneous misfolded glycoprotein interleukin-8 (IL-8) for the elucidation of the recognition pattern by UGGT.[32] In this research, the author chose IL-8 as a synthetic target due to its well-studied tertiary structure, its potential for disulfide shuffling, and the established synthetic strategy. The Asn36, which is not N-glycosylated in nature, was used for incorporation of a mannose-9-high-mannose type N-glycan (M9-glycan). The chemical synthesis of glycosylated IL-8 **96** was achieved by NCL between peptide **94** (1–33) and glycopeptide **95** (34–72), which were prepared by SPPS (Scheme 14).

In order to obtain the intentionally misfolded glycoprotein IL-8 for the investigation of UGGT recognition, the full-length IL-8 **96** was treated under oxidative folding conditions in the absence of redox reagents (cysteine and cystine). As monitored by LC–MS, four products were obtained that were assigned to be native structure **97**, misfolded structures **98** and **99**, and a misfolded dimer **100**. (Scheme 14) With correctly folded and misfolded glycoproteins in hand, finally, the binding assays with UGGT were examined and monitored with LC–MS by addition of one glucose residue. The results indicated that misfolded dimer **100** showed the highest glucosyl transfer activity with UGGT, while the misfolded glycoproteins **98** and **99** showed moderate glucosyl transfer activity relative to the folded glycoprotein **97** as we expected. These results were also in accord with the order of protein surface hydrophobicity (dimer > misfolded > folded).[32] In this research, the authors successfully investigated the recognition pattern of UGGT by using chemically synthesized, homogeneous, intentionally misfolded glycoproteins.

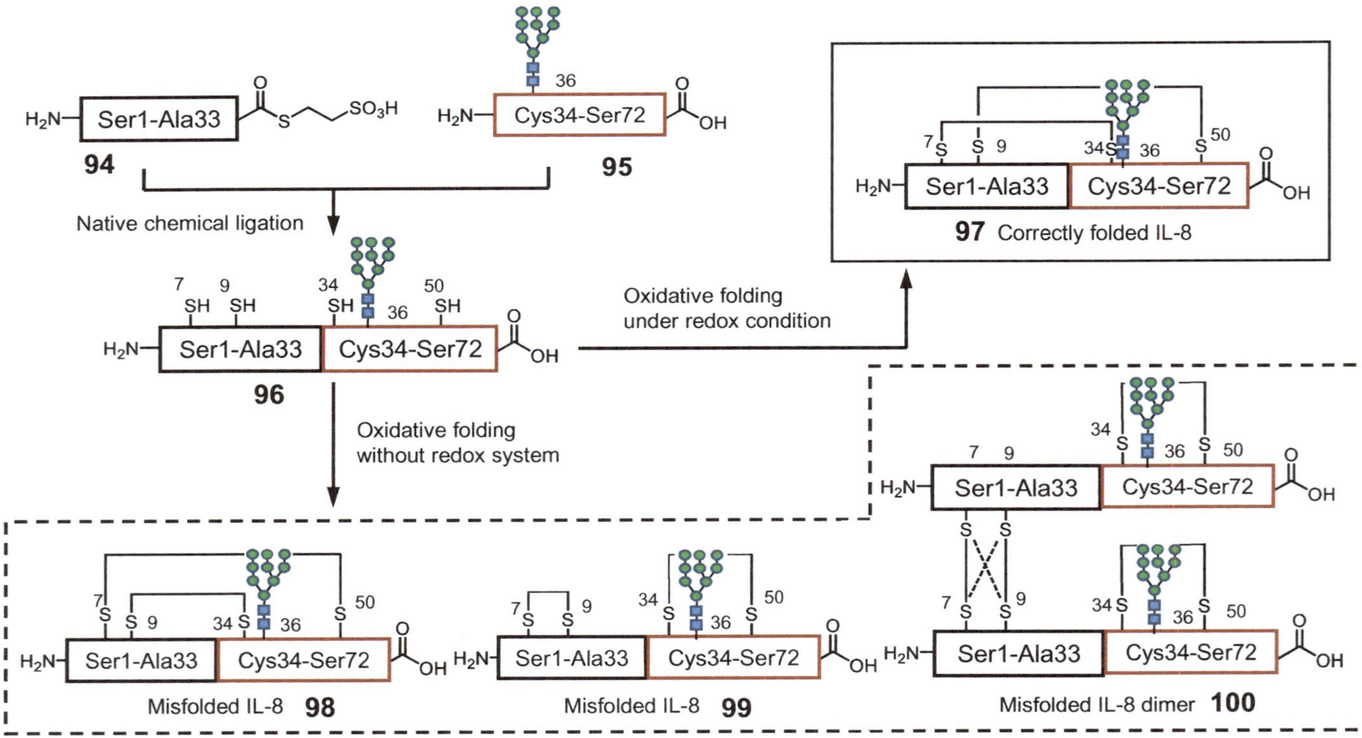

Scheme 14 Chemical synthesis of the homogeneous folded and misfolded N-glycosylated IL-8 for investigation of UGGT recognition.

In order to better understand whether UGGT also recognizes and facilitates the refolding of folding intermediates, based on previous research, the Kajihara group synthesized the N-glycosylated protein crambin bearing an M9-glycan for investigation of UGGT's recognition.[33] Glycosyl crambin was a suitable substrate for this research because the folding process only yielded correctly folded glycoprotein as a sole product. The sequence of crambin was divided into two segments: (1) segment **101** (sequence:1–15) and (2) segment **102** (sequence:16–46, glycosylation: *N*17). After using SPPS to obtain both segments, NCL afforded a full-length glycosylated crambin **103** bearing an M9-glycan that is folded to give native glycosyl crambin **104**. To prepare misfolded crambin **105** and an unproductive intermediate, the native Cys3 or Cys4 was mutated into Ser to hamper disulfide formation. Air oxidation folding in the absence of redox reagents afforded three kinds of misfolded glycosylated crambin **105** through formation of unproductive intermediates. (Scheme 15, top)

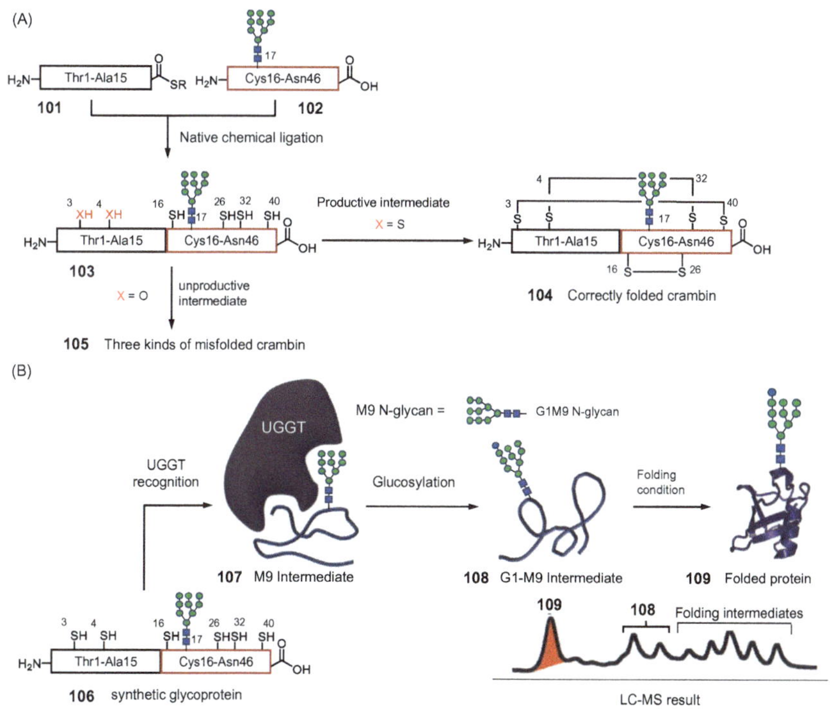

Scheme 15 Chemical synthesis of homogeneous folded and misfolded glycoproteins. (A) Productive (X = S) /nonproductive (X = O) intermediates of N-glycosylated crambin for investigation of UGGT recognition. (B) UGGT assays showed that all intermediates were glucosylated by UGGT.

With the correct native sequence and the mutated sequence M9-glycosyl crambin (X = O) in hand, the UGGT binding assay was carried out using these substrates that yielded G1M9-glycoprotein as product **109**. The result was analyzed by LC–MS with monitoring according to the MS intensity of the G1M9-glycoprotein (Scheme 15B). According to the results, both the productive folding intermediates during the correct folding of the native sequence, and the unproductive intermediates during the misfolding of the mutated sequence were recognized by UGGT to afford the G1M9 products. In this study, the authors were able to elucidate the recognition pattern of UGGT toward folding intermediates.

By using chemically synthesized homogeneous glycoproteins in both correctly folded form and misfolded form, the UGGT recognition pattern, as well as the glycoprotein quality control system in detail, were systematically investigated. In 2018, the Kajihara group synthesized four glycoforms of EPO in both correctly folded and misfolded forms.[34] The chemical synthesis was achieved by SPPS toward both glycopeptides and non-glycosylated peptides, followed with sequential NCL to afford full-length EPOs bearing three M9-glycans at Asn24, 38 and 83. The in vitro folding under dialysis and redox conditions afford both correctly folded and heterogeneous misfolded EPOs **110**. The four pairs of folded/misfolded EPO glycoforms were used for UGGT recognition assays (Scheme 16). However, unexpectedly, all these glycoproteins were shown to be recognized by UGGT to give G1M9 glucosylated product **112**. The authors hypothesized that UGGT might recognize the hydrophobic area on the protein surface and could not distinguish between the folded/misfolded characteristics.

In order to better understand how the refolding of misfolded glycoprotein proceeded in the glycoprotein quality control (GQC) system in the ER, the synthetic misfolded glycoproteins EPO **110** and IL-8 **111** were treated with isolated ER lysate containing all GQC enzymes and chaperons (Scheme 16, bottom right). These assays determined that a fast glucosylation/deglucosylation process exists during the GQC process, and the folding intermediate G1M9-misfolded glycoprotein was efficiently converted to native M9 IL-8 **113**. This research demonstrated that synthetic homogeneous misfolded glycoprotein can be used as a powerful tool for investigating the GQC process in detail.

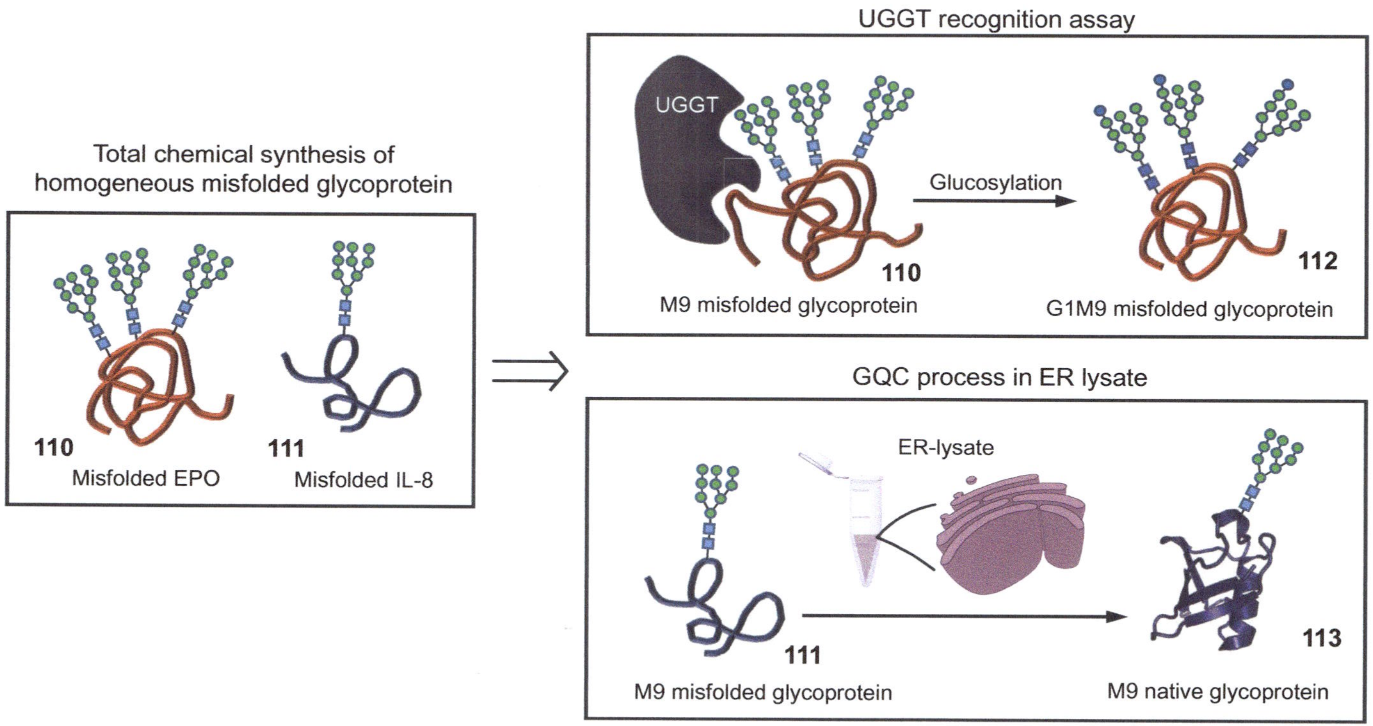

Scheme 16 Chemical synthesis of homogenous glycoproteins for elucidation of UGGT recognition and the GQC process.

5. Semi-synthesis of homogeneous glycoproteins using novel expressed peptide thioesterification methods

In order to solve the problems encountered during the chemical synthesis of glycoproteins, such as the waste of N-glycans and low efficiency of long-chain peptide synthesis, the semi-synthetic method—a process that combines cell expression (usually *E. coli* expression) and chemical synthesis (mainly SPPS)—was developed. *E. coli* expression can produce peptides without glycans up to over 200 residues, while SPPS can produce short glycopeptides with a homogeneous glycan structure. Subsequent ligation between the peptides prepared with *E. coli* and the glycopeptides from SPPS can give full-length homogeneous glycopeptides in only a few conversion steps.

During the semi-synthesis of (glyco)proteins, the main challenge is the thioesterification of peptides obtained from *E. coli* expression, namely expressed peptide thioesterification.[35,36] Expressed peptide thioesterification requires selective C-terminal activation in the absence of any side-chain protection (Scheme 17). The thioesterification methods use such as the Intein system[35] or specific proteases for the recombinant polypeptides[37–39]; however, these enzymatic methods require specific amino acid sequences. On the other hand, Okamoto and coworkers developed reactions for the chemical thioesterification of recombinant peptides by using unique strategies for both C-terminal and internal Cys modifications, In 2012 they reported the internal Cys modification, by S-thiocarbonylation (Scheme 17A).[40] They designed a new thioester synthesis consisting of three steps: (1) The selective S-thiocarbonylation of a β-thiol group of a cysteine residue to provide a phenyl xanthate group by using *O*-phenylchlorothionoformate. (2) Treatment of the resultant peptide **115** with *N*-acetylguanidine to give the peptide thioester surrogate **116**. (3) Finally, conversion of *N*-acetylguanidine derivative **116** into the desired thioester **117**. They also found that *N*-Boc protection of both amino groups of the Lys side chains, and the N-terminus are necessary to avoid side reaction by the *N*-acetylguanidine treatment. In addition to this efficient reaction, peptidyl *N*-acetylguanidine is much less reactive for NCL than the corresponding thioester and can be directly used for a one-pot sequential NCL. Using their thioesterification approach, they succeeded in the semi-synthesis of interleukin 13, which is introduced in the passages that follow (Scheme 18).

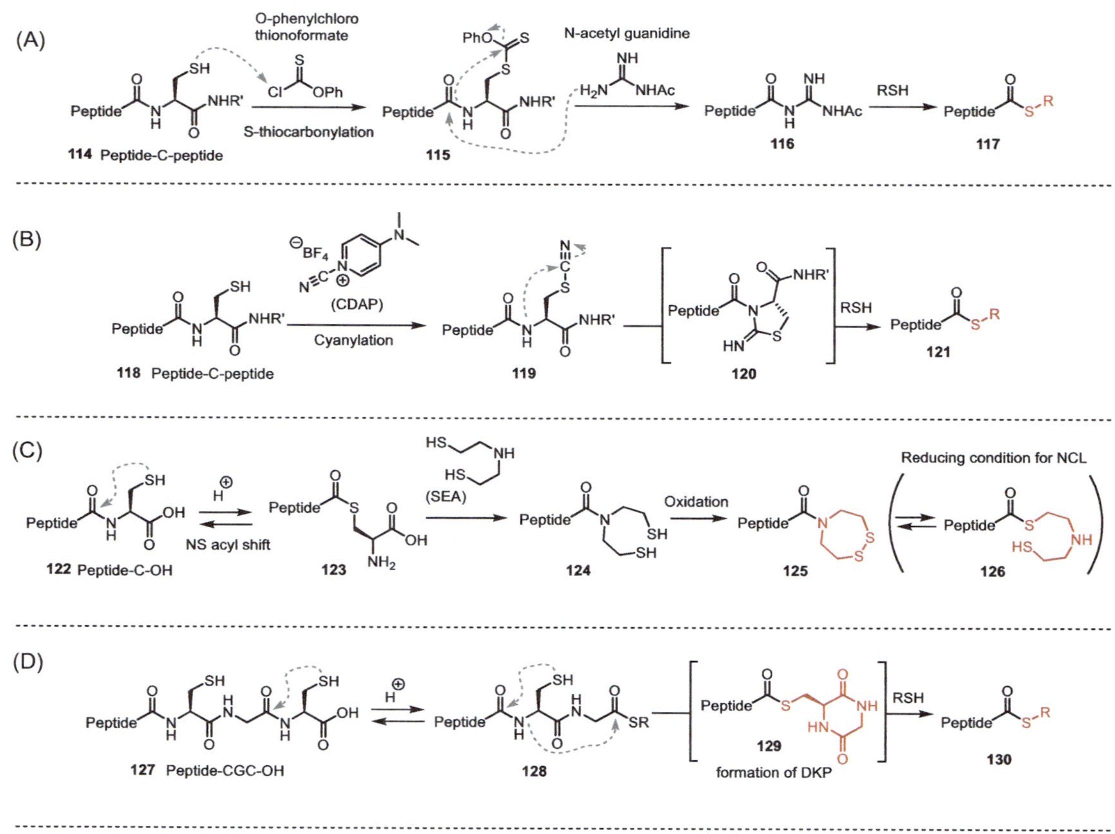

Scheme 17 Cys-based peptide α-thioesterification reactions developed by Kajihara and coworkers. (A) Peptide thioesterification using S-thiocarbonylation, the "guanidine method" (B) Peptide thioesterification by cyanylation. (C) C-terminal Cys-selective thioesterification. (D) C-terminal Cys-selective thioesterification by applying a peptide-CGC sequence.

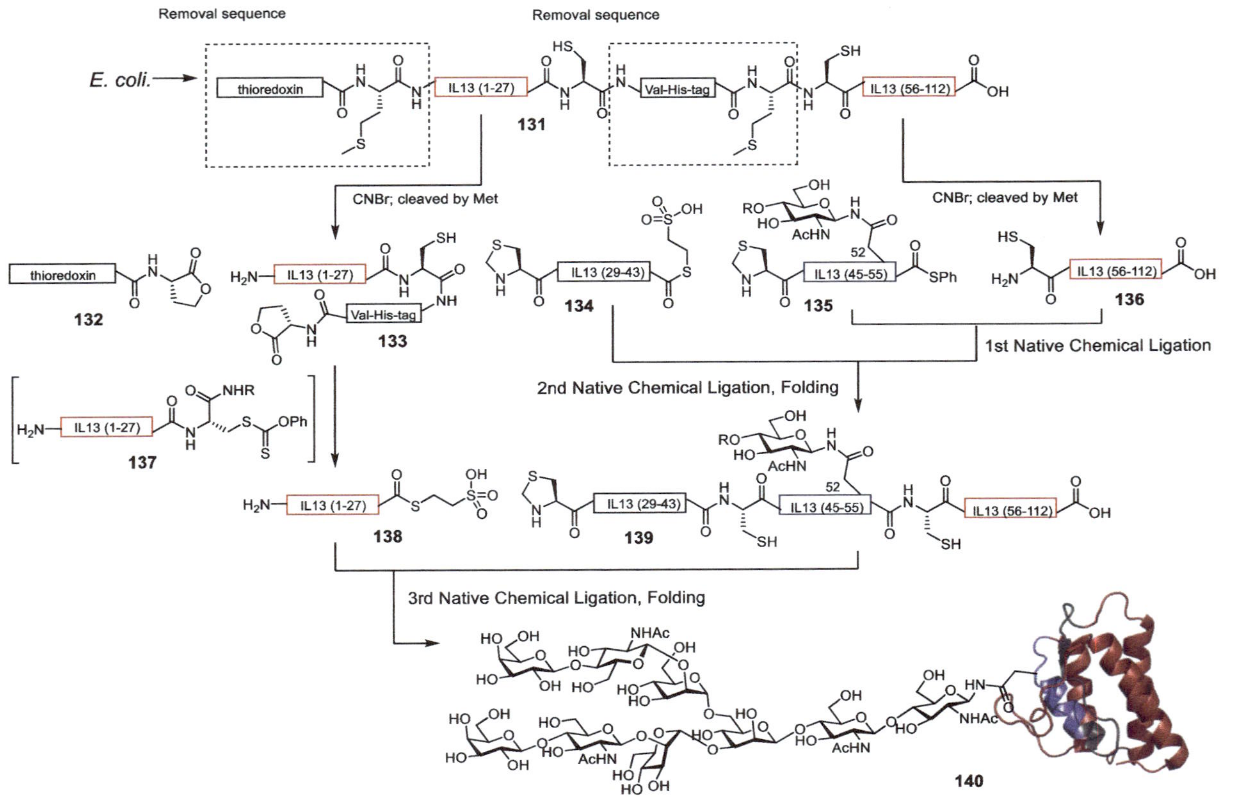

Scheme 18 Semi-synthesis of IL13 using the chemically expressed thioesterification method. The key feature of the semi-synthesis is the simultaneous synthesis of the two peptide segments.

Two years later, Kajihara and coworkers developed another thioesterification method by using cyanylation of the Cys residue.[41] The reaction mechanism is shown in Scheme 17B. First, the thiol group of the cysteine residue is cyanylated by cyanylation reagents. This intermediate **119** is comparably stable, and in the presence of nucleophiles such as hydrazine or amines, the amide nitrogen of the cysteine nucleophilically attacks the sp carbon atom of the cyano group, subsequently forming a five-membered ring intermediate, which efficiently gives the corresponding peptide hydrazide as the thioester surrogate. In their reports, the authors demonstrated the preparation of a polypeptide thioester consisting of 94 amino acids by *E. coli* expression, followed by cyanylation, hydrazinolysis, and thiolysis. Their attempts to synthesize expressed peptide thioesters utilizing internal Cys residues succeeded in synthesizing several homogeneous glycoproteins. However, these thioesterification protocols by Cys residues required multiple reaction steps. The selective modification of the C-terminal Cys was also difficult for peptides containing internal Cys residues at the same time.

To solve these problems, the Okamoto and Kajihara group reported two types of highly chemo-selective peptide thioesterification reactions in 2022.[42] These new thioesterification reactions progressed without any protecting groups by applying an intramolecular N—S acyl shift reaction at a C-terminal Cys residue (Scheme 17C and D). These authors made a much-improved thioesterification based on reactions developed by the Aimoto[43,44] and Macmillan[45] groups. The first method is the chemical transformation of peptide-Cys **122** using bis(2-sulfanylethyl)amine (SEA) that efficiently yields peptide-SEA **125**, which can be utilized as a peptide thioester surrogate (Scheme 17C).[42] The reaction starts from conversion of peptide-Cys **122** to peptide-S-Cys-thioester **123** by an intramolecular N—S acyl shift at the C-terminal Cys and the trans-thioesterification of the peptide-S-Cys-thioester **123** with an external thiol such as SEA to afford peptide-SEA **124** (peptide thioester surrogate). Fortunately, the equilibrium state is biased toward peptide-SEA **125** because peptide-SEA **124** is oxidized in air to form C-terminal disulfide bonds, which cannot be converted to peptide-S-Cys-thioester **123**.

Alternatively, these workers also found that a peptide having cysteinyl glycyl cysteine **127** (peptide-CGC) can be used for the synthesis of peptide thioesters from unprotected polypeptides (Scheme 17D).[42] Peptides-CGC **127** could be transformed to peptide-thioesters **130** through peptides with CG-thioester units **128** at the C-terminus via an N—S acyl shift at Cys. Then peptide-CG-thioester **128** can be converted to peptide thioesters

130 via diketopiperazine (DKP) formation and a subsequent N—S acyl shift at Cys. Both methods were successfully applied to recombinant polypeptides and utilized for NCL. In the next paragraph, several examples of the semi-synthesis applying Cys-based thioesterification protocols developed by the Okamoto and Kajihara group will be introduced.

By applying these recently developed thioesterification approaches, the Okamoto and Kajihara grouphave reported the synthesis of interleukin 13 (IL-13) in 2014 (Scheme 18) and the inducible T cell co-stimulator (ICOS) in 2022 (Scheme 19).[42,46] At first they reported a new semi-synthetic strategy for IL-13 in which the middle region is glycosylated.[46] IL-13 is a glycoprotein that consists of 112 amino acids and has a complex-type oligosaccharide at Asn52. Using their synthetic strategy for IL-13, they employed four peptide segments and three sequential NCLs to obtain a full-length glycoprotein. Then the IL-13 segment **133** (1–27) and **136** (55–112) were simultaneously prepared by the *E. coli.* expression system and then cleaved fusion proteins **131** by CNBr, which can selectively cleave an amide bond between the Met and Cys residues; furthermore, the Met residue is converted into the homoserine lactone **133** (Scheme 18). They utilized a cleavage reaction on the amide bond between artificially inserted Met and Cys with CNBr to afford segment **133** (1–27) and **136** (55–112) simultaneously and in high yield. Then segment **133** (1–27) was converted to peptide thioesters **138** by using their developed "guanidine method" (Scheme 17A). On the other hand, glycopeptide thioester **135** was synthesized by Fmoc-standard SPPS protocol, followed by their reported protocol.[23] After preparation of all segments by semi-synthetic approaches, sequential NCL was convergently progressed to afford the linear full-length IL13 segment. The folding process under stepwise dialysis with redox reagents yielded folded IL13 **140**. ELISA and a CD spectrum indicated that semi-synthesized IL13 correctly folded and had an α-helix-rich conformation.

In the semi-synthesis of ICOS, the Okamoto and Kajihara group applied peptide-Cys thioesterification by using SEA (Scheme 19).[42] ICOS is a co-stimulator for the activation of T cells and has N-glycosylation sites in the extracellular domain. The sequence of ICOS consists of 106 amino acids including two biantennary sialylglycans at Asn 89 and Asn110. A convergent synthetic strategy was established for glycosyl ICOS using a recombinant polypeptide thioesterification as a key reaction. These workers employed three peptide segments and two sequential NCL to obtain a full-length glycoprotein. ICOS segment **141** (1–62) was co-expressed with the Small

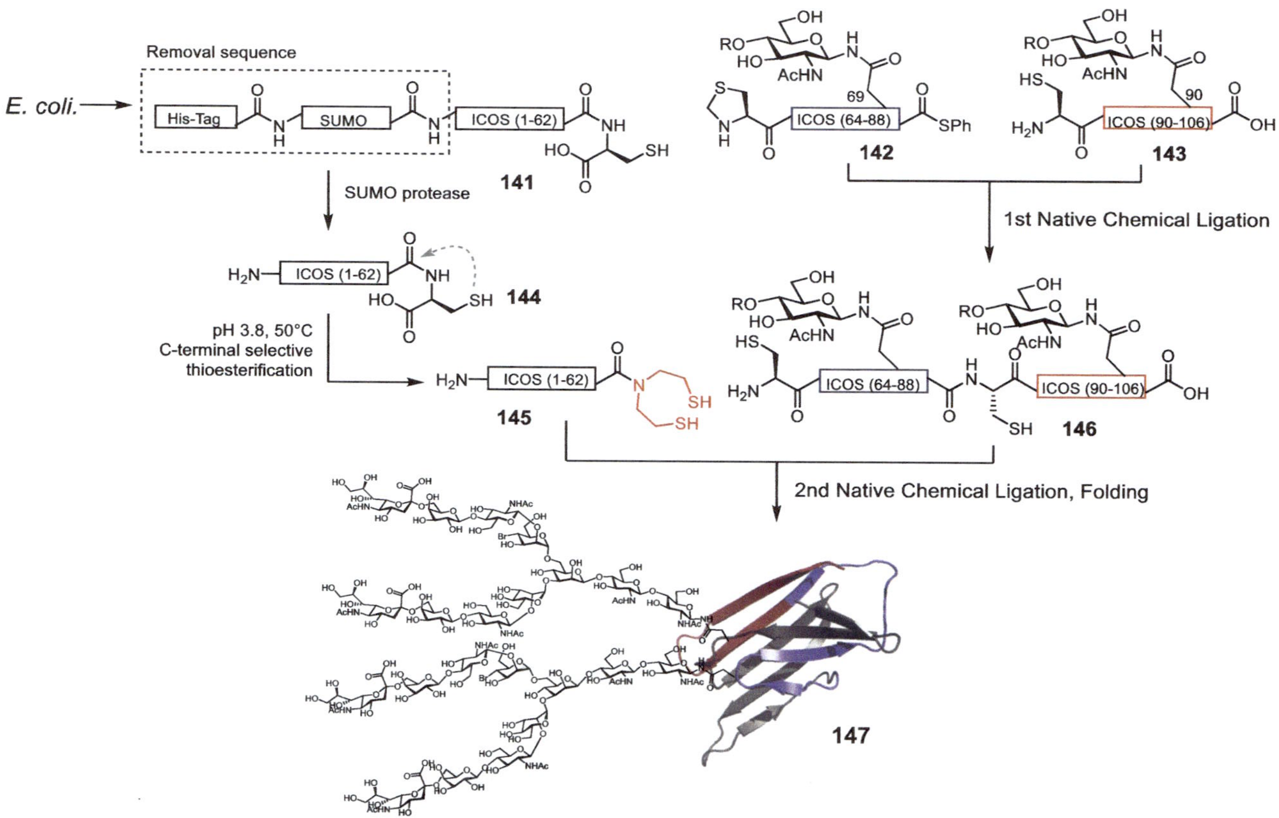

Scheme 19 Semi-synthesis of ICOS by utilizing a novel thioesterification approach via peptide-SEA.

Ubiquitin-related Modifier (SUMO) and six His (His Tag) sequences by an *E. coli* expression system. Thioesterification-applied SEA was progressed convergently after cleavage of the SUMO sequences by SUMO protease to afford peptide-SEA **145**. The peptide-SEA **145** that was obtained was subjected to NCL with glycopeptide **146** prepared from glycopeptide thioester **142** (63–88) and cysteinyl peptide **143** (89–106) synthesized by a standard Boc-SPPS protocol. In terms of glycopeptide-thioester **142**, they employed Fmoc-SPPS. The resultant product from all the segments was sequentially coupled by NCL to obtain a full-length glycopeptide of ICOS. Using stepwise dialysis conditions, the folding process with redox reagents yielded folded glycosyl ICOS **147** (Scheme 19). ELISA, CD and the MS spectrum indicated that semi-synthesized ICOS was the correctly folded structure having a β-sheet rich conformation. By using both chemical synthesis and semi-synthesis as appropriate, it might be possible to synthesize glycoproteins with larger molecular weights.

In conclusion, by combining *E. coli* expression and chemical synthesis, the semi-synthesis protocol provides another option for obtaining homogeneous glycoproteins. The synthetic efficiency is improved due to the facile preparation of a non-glycosylated, long peptide by *E. coli* expression instead of by SPPS. The full-length glycoprotein can be assembled in only few chemical conversion steps, which enhances the total yield and saves the precious N-glycan part during synthesis. Especially in recently reported research, a robust synthetic strategy based on thioacid derivatives without using SPPS (Scheme 13) have been established. The total yield is greatly enhanced compared with the total chemical synthetic method toward glycoproteins. The main challenge remains for the semi-synthesis of large-size glycoproteins, which is research in progress.

Reference

1. Walsh, G.; Jefferis, R. Post-Translational Modifications in the Context of Therapeutic Proteins. *Nat. Biotechnol.* **2006**, *24*, 1241–1252.
2. Okayama, A.; Miyagi, Y.; Oshita, F.; Ito, H.; Nakayama, H.; Nishi, M.; Kurata, Y.; Kimura, Y.; Ryo, A.; Hirano, H. Identification of Tyrosine-Phosphorylated Proteins Upregulated During Epithelial–Mesenchymal Transition Induced With TGF-β. *J. Proteome Res.* **2015**, *14*, 4127–4136.
3. Nagata, K.; Kawakami, T.; Kurata, Y.; Kimura, Y.; Suzuki, Y.; Nagata, T.; Sakuma, Y.; Miyagi, Y.; Hirano, H. Augmentation of Multiple Protein Kinase Activities Associated with Secondary Imatinib Resistance in Gastrointestinal Stromal Tumors as Revealed by Quantitative Phosphoproteome Analysis. *J. Proteomics* **2015**, *115*, 132–142.
4. Ishigami, A.; Masutomi, H.; Handa, S.; Nakamura, M.; Nakaya, S.; Uchida, Y.; Saito, Y.; Murayama, S.; Jang, B.; Jeon, Y. C.; Choi, E. K.; Kim, Y. S.; Kasahara, Y.; Maruyama, N.; Toda, T. Mass Spectrometric Identification of Citrullination Sites

and Immunohistochemical Detection of Citrullinated Glial Fibrillary Acidic Protein in Alzheimer's Disease Brains. *J. Neurosci. Res.* **2015**, *93*, 1664–1674.
5. Ishigami, A.; Ohsawa, T.; Hiratsuka, M.; Taguchi, H.; Kobayashi, S.; Saito, Y.; Murayama, S.; Asaga, H.; Toda, T.; Kimura, N.; Maruyama, N. Abnormal Accumulation of Citrullinated Proteins Catalyzed by Peptidylarginine Deiminase in Hippocampal Extracts From Patients With Alzheimer's Disease. *J. Neurosci. Res.* **2005**, *80*, 120–128.
6. Wong, C.-H. Protein Glycosylation: New Challenges and Opportunities. *J. Org. Chem.* **2005**, *70*, 4219–4225.
7. Unverzagt, C.; Kajihara, Y. Recent Advances in the Chemical Synthesis of N-linked Glycoproteins. *Curr. Opin. Chem. Biol.* **2018**, *46*, 130–137.
8. Agouridas, V.; El Mahdi, O.; Diemer, V.; Cargoet, M.; Monbaliu, J. M.; Melnyk, O. Native Chemical Ligation and Extended Methods: Mechanisms, Catalysis, Scope, and Limitations. *Chem. Rev.* **2019**, *119*, 7328–7443.
9. Dawson, P. E.; Kent, S. B. Synthesis of Native Proteins by Chemical Ligation. *Annu. Rev. Biochem.* **2000**, *69*, 923–960.
10. Dawson, P. E.; Muir, T. W.; Clark-Lewis, I.; Kent, S. B. Synthesis of Proteins by Native Chemical Ligation. *Science* **1994**, *266*, 776–779.
11. Yan, L. Z.; Dawson, P. E. Synthesis of Peptides and Proteins Without Cysteine Residues by Native Chemical Ligation Combined With Desulfurization. *J. Am. Chem. Soc.* **2001**, *123*, 526–533.
12. Wan, Q.; Danishefsky, S. J. Free-Radical-Based, Specific Desulfurization of Cysteine: A Powerful Advance in the Synthesis of Polypeptides and Glycopolypeptides. *Angew. Chem. Int. Ed.* **2007**, *46*, 9248–9252.
13. Thompson, R. E.; Chan, B.; Radom, L.; Jolliffe, K. A.; Payne, R. J. Chemoselective Peptide Ligation–Desulfurization at Aspartate. *Angew. Chem. Int. Ed.* **2013**, *52*, 9723–9727.
14. Kulkarni, S. S.; Sayers, J.; Premdjee, B.; Payne, R. J. Rapid and Efficient Protein Synthesis Through Expansion of the Native Chemical Ligation Concept. *Nat. Rev. Chem.* **2018**, *2*, 0122.
15. Wong, C. T. T.; Tung, C. L.; Li, X. Synthetic Cysteine Surrogates Used in Native Chemical Ligation. *Mol. Biosyst.* **2013**, *9*, 826–833.
16. Payne, R. J.; Wong, C. H. Advances in Chemical Ligation Strategies for the Synthesis of Glycopeptides and Glycoproteins. *Chem. Commun. (Cambridge)* **2010**, *46*, 21–43.
17. Seko, A.; Koketsu, M.; Nishizono, M.; Enoki, Y.; Ibrahim, H. R.; Juneja, L. R.; Kim, M.; Yamamoto, T. Occurrence of a Sialylglycopeptide and Free Sialylglycans in Hen's Egg Yolk. *Biochim. Biophys. Acta.* **1997**, *1335* (1-2), 23–32.
18. Fukae, K.; Yamamoto, N.; Hatakeyama, Y.; Kajihara, Y. Chemoenzymatic Synthesis of Diverse Asparagine-Linked Alpha-(2,3)-sialyloligosaccharides. *Glycoconjugate J.* **2004**, *21*, 243–250.
19. Kajihara, Y.; Suzuki, Y.; Yamamoto, N.; Sasaki, K.; Sakakibara, T.; Juneja, L. R. Prompt Chemoenzymatic Synthesis of Diverse Complex-Type Oligosaccharides and Its Application to the Solid-Phase Synthesis of a Glycopeptide With Asn-Linked Sialyl-undeca- and Asialo-Nonasaccharides. *Chem. Eur. J.* **2004**, *10*, 971–985.
20. Maki, Y.; Okamoto, R.; Izumi, M.; Murase, T.; Kajihara, Y. Semisynthesis of Intact Complex-Type Triantennary Oligosaccharides From a Biantennary Oligosaccharide Isolated From a Natural Source by Selective Chemical and Enzymatic Glycosylation. *J. Am. Chem. Soc.* **2016**, *138*, 3461–3468.
21. Maki, Y.; Okamoto, R.; Izumi, M.; et al. Chemical Synthesis of an Erythropoietin Glycoform Having a Triantennary N-Glycan: Significant Change of Biological Activity of Glycoprotein by Addition of a Small Molecular Weight Trisaccharide. *J. Am. Chem. Soc.* **2020**, *142*, 20671–20679.

22. Yamamoto, N.; Takayanagi, A.; Yoshino, A.; Sakakibara, T.; Kajihara, Y. An Approach for a Synthesis of Asparagine-Linked Sialylglycopeptides Having Intact and Homogeneous Complex-Type Undecadisialyloligosaccharides. *Chem. Eur. J.* **2007**, *13*, 613–625.
23. Yamamoto, N.; Tanabe, Y.; Okamoto, R.; Dawson, P. E.; Kajihara, Y. Chemical Synthesis of a Glycoprotein Having an Intact Human Complex-Type Sialyloligosaccharide under the Boc and Fmoc Synthetic Strategies. *J. Am. Chem. Soc.* **2008**, *130*, 501–510.
24. Sakamoto, I.; Tezuka, K.; Fukae, K.; Ishii, K.; Taduru, K.; Maeda, M.; Ouchi, M.; Yoshida, K.; Nambu, Y.; Igarashi, J.; Hayashi, N.; Tsuji, T.; Kajihara, Y. Chemical Synthesis of Homogeneous Human Glycosyl-Interferon-β That Exhibits Potent Antitumor Activity In Vivo. *J. Am. Chem. Soc.* **2012**, *134*, 5428–5431.
25. Murakami, M.; Okamoto, R.; Izumi, M.; Kajihara, Y. Chemical Synthesis of an Erythropoietin Glycoform Containing a Complex-Type Disialyloligosaccharide. *Angew. Chem. Int. Ed.* **2012**, *51*, 3567–3572.
26. Murakami, M.; Kiuchi, T.; Nishihara, M.; Tezuka, K.; Okamoto, R.; Izumi, M.; Kajihara, Y. Chemical Synthesis of Erythropoietin Glycoforms for Insights Into the Relationship Between Glycosylation Pattern and Bioactivity. *Sci. Adv.* **2016**, *2*, e1500678.
27. Okamoto, R.; Mandal, K.; Ling, M.; Luster, A. D.; Kajihara, Y.; Kent, S. B. Total Chemical Synthesis and Biological Activities of Glycosylated and Non-glycosylated Forms of the Chemokines CCL1 and Ser-CCL1. *Angew. Chem. Int. Ed.* **2014**, *53*, 5188–5193.
28. Minh Hien, N.; Izumi, M.; Sato, H.; Okamoto, R.; Kajihara, Y. Chemical Synthesis of Glycoproteins With the Specific Installation of Gradient-Enriched ^{15}N-Labeled Amino Acids for Getting Insights into Glycoprotein Behavior. *Chem. Eur. J.* **2017**, *23*, 6579–6585.
29. Nomura, K.; Maki, Y.; Okamoto, R.; Satoh, A.; Kajihara, Y. Glycoprotein Semisynthesis by Chemical Insertion of Glycosyl Asparagine Using a Bifunctional Thioacid-Mediated Strategy. *J. Am. Chem. Soc.* **2021**, *143*, 10157–10167.
30. Tam, J. P.; Lu, Y. A.; Liu, C. F.; Shao, J. Peptide Synthesis Using Unprotected Peptides Through Orthogonal Coupling Methods. *Proc. Natl. Acad. Sci. U. S. A.* **1995**, *92*, 12485–12489.
31. Zhang, X.; Li, F.; Liu, C. F. Synthesis of Histone H3 Proteins by a Thioacid Capture Ligation Strategy. *Chem. Commun. (Cambridge)* **2011**, *47*, 1746–1748.
32. Izumi, M.; Makimura, Y.; Dedola, S.; Seko, A.; Kanamori, A.; Sakono, M.; Ito, Y.; Kajihara, Y. Chemical Synthesis of Intentionally Misfolded Homogeneous Glycoprotein: A Unique Approach for the Study of Glycoprotein Quality Control. *J. Am. Chem. Soc.* **2012**, *134*, 7238–7241.
33. Dedola, S.; Izumi, M.; Makimura, Y.; Seko, A.; Kanamori, A.; Sakono, M.; Ito, Y.; Kajihara, Y. Folding of Synthetic Homogeneous Glycoproteins in the Presence of a Glycoprotein Folding Sensor Enzyme. *Angew. Chem. Int. Ed.* **2014**, *53*, 2883–2887.
34. Kiuchi, T.; Izumi, M.; Mukogawa, Y.; Shimada, A.; Okamoto, R.; Seko, A.; Sakono, M.; Takeda, Y.; Ito, Y.; Kajihara, Y. Monitoring of Glycoprotein Quality Control System With a Series of Chemically Synthesized Homogeneous Native and Misfolded Glycoproteins. *J. Am. Chem. Soc.* **2018**, *140*, 17499–17507.
35. Muir, T. W.; Sondhi, D.; Cole, P. A. Expressed Protein Ligation: A General Method for Protein Engineering. *Proc. Natl. Acad. Sci. U. S. A.* **1998**, *95*, 6705–6710.
36. Muir, T. W. Semisynthesis of Proteins by Expressed Protein Ligation. *Annu. Rev. Biochem.* **2003**, *72*, 249–289.
37. Ling, J. J.; Policarpo, R. L.; Rabideau, A. E.; Liao, X.; Pentelute, B. L. Protein Thioester Synthesis Enabled by Sortase. *J. Am. Chem. Soc.* **2012**, *134*, 10749–10752.

38. Nguyen, G. K.; Wang, S.; Qiu, Y.; Hemu, X.; Lian, Y.; Tam, J. P. Butelase 1 is an Asx-specific Ligase Enabling Peptide Macrocyclization and Synthesis. *Nat. Chem. Biol.* **2014**, *10*, 732–738.
39. Komiya, C.; Shigenaga, A.; Tsukimoto, J.; Ueda, M.; Morisaki, T.; Inokuma, T.; Itoh, K.; Otaka, A. Traceless Synthesis of Protein Thioesters Using Enzyme-Mediated Hydrazinolysis and Subsequent Self-Editing of the Cysteinyl Prolyl Sequence. *Chem. Commun.(Cambridge)* **2019**, *55*, 7029–7032.
40. Okamoto, R.; Morooka, K.; Kajihara, Y. A Synthetic Approach to a Peptide α-Thioester From an Unprotected Peptide Through Cleavage and Activation of a Specific Peptide Bond by *N*-acetylguanidine. *Angew. Chem. Int. Ed.* **2012**, *51*, 191–196.
41. Kajihara, Y.; Kanemitsu, Y.; Nishihara, M.; Okamoto, R.; Izumi, M. Efficient Synthesis of Polypeptide-α-thioester by the Method Combining Polypeptide Expression and Chemical Activation for the Semi-Synthesis of Interferon-γ Having Oligosaccharides. *J. Pept. Sci.* **2014**, *20*, 958–963.
42. Okamoto, R.; Iritani, K.; Amazaki, Y.; Zhao, D.; Chandrashekar, C.; Maki, Y.; Kanemitsu, Y.; Kaino, T.; Kajihara, Y. Semisynthesis of a Homogeneous Glycoprotein Using Chemical Transformation of Peptides to Thioester Surrogates. *J. Org. Chem.* **2022**, *87*, 114–124.
43. Kawakami, T.; Sumida, M.; Nakamura, K. I.; Vorherr, T.; Aimoto, S. Peptide Thioester Preparation Based on an N-S Acyl Shift Reaction Mediated by a Thiol Ligation Auxiliary. *Tetrahedron Lett.* **2005**, *46*, 8805–8807.
44. Nakamura, K. I.; Sumida, M.; Kawakami, T.; Vorherr, T.; Aimoto, S. Generation of an S-Peptide via an N–S Acyl Shift Reaction in a TFA Solution. *Bull. Chem. Soc. Jpn.* **2006**, *79*, 1773–1780.
45. Kang, J.; Macmillan, D. Peptide and Protein Thioester Synthesis via N→S Acyl Transfer. *Org. Biomol. Chem.* **2010**, *8*, 1993–2002.
46. Okamoto, R.; Kimura, M.; Ishimizu, T.; Izumi, M.; Kajihara, Y. Semisynthesis of a Post-Translationally Modified Protein by Using Chemical Cleavage and Activation of an Expressed Fusion Polypeptide. *Chem. Eur. J.* **2014**, *20*, 10425–10430.

CHAPTER FIVE

A bacterial glycolipid essential for membrane protein integration☆

Kohki Fujikawa[a], Shoko Mori[a,b], Ken-ichi Nishiyama[c], and Keiko Shimamoto[a,b,*]

[a]Bioorganic Research Institute, Suntory Foundation for Life Sciences, Kyoto, Japan
[b]Department of Chemistry, Graduate School of Science, Osaka University, Toyonaka, Japan
[c]Department of Biological Chemistry and Food Sciences, Faculty of Agriculture, Iwate University, Morioka, Japan
*Corresponding author: e-mail address: shimamot@sunbor.or.jp

Contents

☆Dedicated to the memory of Professor Hidetoshi Yamada

Advances in Carbohydrate Chemistry and Biochemistry, Volume 81
ISSN 0065-2318
https://doi.org/10.1016/bs.accb.2022.09.005

Abbreviations

Ac	acetyl
Bn	benzyl
Bz	benzoyl
CD	circular dichroism
CDP	cytidine diphosphate
DAG	diacylglycerol
E. coli	*Escherichia coli*
ECA	enterobacterial common antigen
EPL	*E. coli* phospholipids
ER	endoplasmic reticulum
Fuc4NAc	*N*-acetyl-4-amino-4-deoxy-D-fucose
GC	gas chromatography
GlcNAc	*N*-acetyl-glucosamine; 2-Acetamido-2-deoxy-D-glucose
INV	inverted membrane vesicles
K_D	dissociation constant
LPS	lipopolysaccharide
MALDI-TOF-MS	matrix-assisted laser desorption/ionization-time of flight-mass spectrometry
ManNAcA	*N*-acetylmannosaminuronate; 2-acetamido-2-deoxy-D-mannuronic acid
MeCN	acetonitrile
MPIase	membrane protein integrase
MS/MS	tandem mass spectrometry
MtlA	mannitol permease
NC	nascent chain
NMR	nuclear magnetic resonance
NOESY	nuclear Overhauser effect spectroscopy
PA	phosphatidic acid
PMB	*p*-methoxybenzyl
PMF	proton motive force
pOmpA	precursor form of outer membrane protein A
Q-TOF-MS	quadrupole-time-of-flight-mass spectrometry
SDS-PAGE	sodium dodecyl sulfate–polyacrylamide gel electrophoresis
SPR	surface plasmon resonance
SR	signal recognition particle receptor
SRP	signal recognition particle
STD	saturation transfer difference
TA protein	tail-anchored protein
TBDPS	*tert*-butyldiphenylsilyl
Tf	trifluoromethanesulfonyl (triflyl)
TLC	thin-layer chromatography
TM	transmembrane

1. Introduction

Numerous proteins situated in the biomembranes play a vital role in cells. These membrane proteins have hydrophobic transmembrane (TM) regions. Since they are newly synthesized in the ribosomes in a cytosolic aqueous environment, integrating the nascent chains (NCs) into membranes ensures proper conformation and orientation. This mechanism is similar to the signal hypothesis revealed by Blobel,[1] which describes how proteins are destined for the endoplasmic reticulum (ER) or other organelles.[2] A cleavable signal peptide tag (16–30 amino acid residues) is present at the N-terminus of the protein targeted to the ER membrane in eukaryotes. The signal-recognition particle (SRP) then recognizes the signal peptide generated from a ribosome and leads it to the SRP receptor (SR) present on the surface of the membrane. Protein complex channels called translocons are used to facilitate the crossing of these proteins in the membrane.[3,4] Immediately after transferring the ribosome-NC complex from the SR to the cytoplasmic face of the translocon, the signal sequence is inserted into the translocon, and the polypeptide chain co-translationally grows through the translocon. A signal peptidase may cleave during or after translocation to generate a mature protein. In the case of the membrane protein integration, its first TM functions instead of the signal peptide, but it is not cleaved. The hydrophobic part of the polypeptide chain threaded from the ribosome laterally extrudes into the lipid bilayer according to the conformational change of the translocon. By contrast, the hydrophilic part is located outside the membrane. Membrane proteins with multiple TM regions form structures by repeating these processes.

The core of the translocon is the Sec heterotrimer (Sec YEG in bacteria and Sec61 in eukaryotic cells, respectively), which functions as a unit forming a pore. Also, the molecular bases of the translocation of secretory proteins and the integration of membrane proteins are highly conserved in all organisms since homologous translocons are generally present in prokaryotes such as *Escherichia coli* (*E. coli*) and eukaryotes such as humans cells.[5] Nevertheless, the translocon efficiently performs various functional roles since it has various partner proteins, and it may oligomerize with a dynamic association in different complexes.

In the membrane protein integration in *E. coli* (Fig. 1A), the translocon comprises SecY, SecE, and SecG. The SRP comprises Ffh and 4.5S RNA, and SR corresponds to FtsY.[6] An additional proteinaceous factor, YidC,

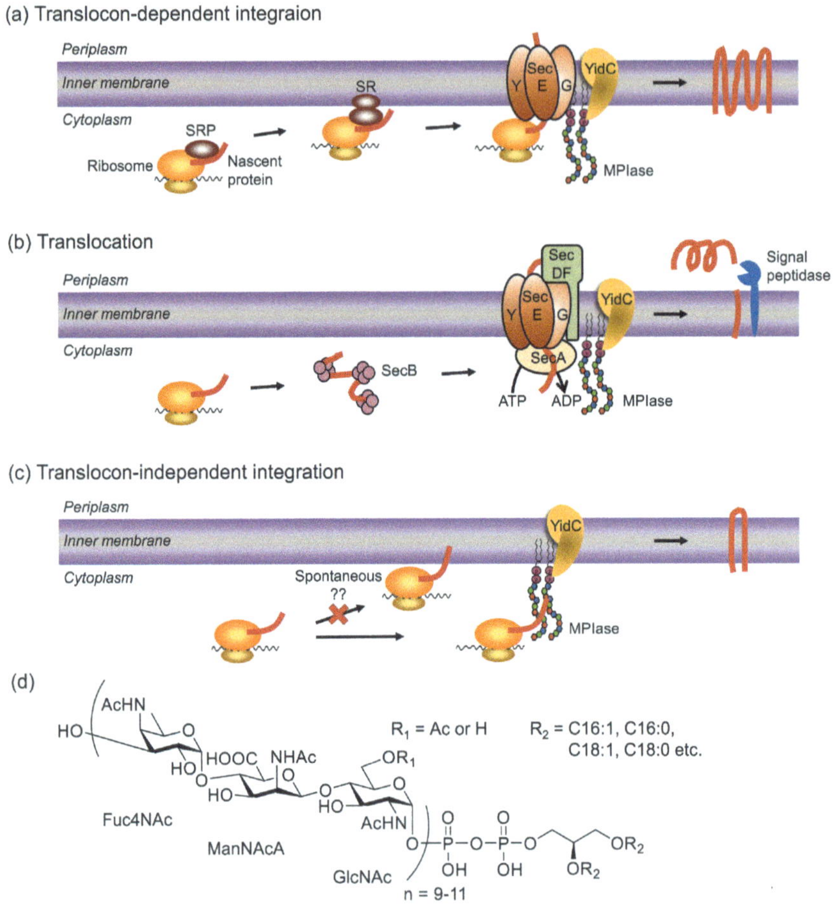

Fig. 1 Schematic model for membrane protein integration and translocation in *E. coli*. MPlase is involved in each pathway, but colocalization of MPlase with SecYEG/YidC is speculative at this stage. (A) Translocon-dependent membrane protein integration pathway. (B) Translocon-dependent translocation pathway. (C) Translocon-independent integration pathway. (D) Structure of natural MPlase.

may function as an insertase for the membrane integration and/or facilitate the lateral movement the TM segments into the bilayer.[7–9] Meanwhile, in the translocation of a less hydrophobic protein intended for the outer membrane in *E. coli* (Fig. 1B), the signal sequence is post-translationally recognized by the molecular chaperone SecB instead of SRP, and the preprotein is led to SecA. The motor protein SecA catalyzes the protein insertion as it hydrolyzes ATP. First, the preprotein penetrates the membrane through the complex consisting of SecYEG, SecDF(yajC), and

YidC. Then, the signal sequence anchored in the membrane is cleaved by the signal peptidase, and the extramembrane region is released.

Additionally, a translocon-independent pathway is known for a subset of small-membrane proteins[10] and tail-anchored (TA) membrane proteins[11] because they cannot bind to the SRP due to the location of their TM region at the C-terminus. Although ribosome-associating proteins serve as chaperones for TA membrane proteins in eukaryotes,[12] such chaperones are unidentified in prokaryotes.[13] Consequently, in *E. coli*, this pathway was previously believed to be spontaneous and facilitated by hydrophobic interactions between the TM regions of the protein and the membrane lipids (Fig. 1C).[14]

However, we identified a novel factor promoting membrane protein integration named MPIase (membrane protein integrase).[15–17] Applying structural analyses, we showed that MPIase is a glycolipid despite its enzyme-like activity (Fig. 1D).[15] In this review, we describe the identification of MPIase and the elucidation of its activity mechanisms using synthetic analogs.

2. Identification of MPIase

2.1 Inhibition of the spontaneous integration

Historically, the effects of translocons were analyzed using the protease protection assay (described below) via the in vitro translational system and inverted membrane vesicles (INVs) prepared from the inner membrane of *E. coli* mutants of which the related factors were genetically modified.[18] Since INV contains various endogenous components, Nishiyama used artificial (proteo)liposomes prepared from *E. coli* phospholipids (EPL) instead of INVs, excluding other factors except for translocons. However, the in vitro translational system using EPL liposomes could not accurately reproduce membrane integration. For instance, mannitol permease (MtlA), a translocon-dependent membrane protein with 6TM regions, spontaneously integrated into EPL liposomes even without the assistance of translocons or YidC. Nevertheless, such spontaneous integration was strictly blocked in INV, suggesting that additional unknown components in INV regulate the integration pathways.[16,19] In an effort to identify them, it was revealed that diacylglycerol (DAG), a minor non-phospholipid component in INV, prevents indiscriminate spontaneous integration in biomembranes.[19,20] DAG is an essential second messenger in eukaryotic cells, but *E. coli* lacks this signaling pathway. The existence of DAG in the bacterial membrane had been shown, however, its precise function in bacteria was unclear. Proteoliposomes containing SecYEG and YidC in the presence of DAG were not active as to MtlA integration. Besides, a physiological concentration of DAG (2–3 w/w%) significantly suppressed the spontaneous integration of the Sec-independent proteins into EPL liposomes.[20] DAGs possessing C8–C18 fatty acids showed similar blocking activities for membrane protein integration. If hydrophobic molecules spontaneously and freely integrate into biomembranes, the biomembrane may no longer function as a barrier. Therefore, it is plausible that biomembranes have a strategy to prevent disordered spontaneous integration.

2.2 Membrane protein integration promoting factor

If spontaneous integration does not occur due to the endogenous DAG in the *E. coli* inner membrane, there must be another way to promote favorable integration of translocon-independent proteins. It was revealed that adding *E. coli* inner membrane extract to DAG-containing liposomes restored membrane protein integration activity, strongly suggesting the presence of an unknown factor promoting the integration.[19] The factor was named "MPIase (membrane protein integrase)" after its function.[15,21]

2.2.1 Purification of MPIase

INVs prepared from *E. coli* MC4100 were washed with 5 M urea and then extracted with sodium cholate.[21] The extract did not contain detectable levels of Sec factors, YidC or SRP after it had been precipitated with acetone and then solubilized in a detergent solution. The extract was active in the integration assay, and the integration activity was further enhanced through several column chromatography separations. Although the amphiphilicity of MPIase hampered the purification, after much trial and error, a slightly contaminated lipopolysaccharide (LPS) was removed by liquid–liquid partition chromatography.[16,21] The column (Sephadex LH-20) was equilibrated with the water phase of the solvent mixture (1-butanol/tetrahydrofuran/ethanol/water: 8/7/1/20) and was developed with its organic phase. We obtained an active fraction showing a 7 kDa band on a urea/SDS-gel, which gave a brown spot in TLC analysis involving a chromogenic reaction with anisaldehyde, while LPS gave a gray zone.[21]

We successfully reconstructed both translocon-dependent[19,22] and -independent[21] membrane integration systems by adding the purified MPIase and DAG into EPL liposomes. Thus, MPIase is essential in both pathways. Additionally, MPIase stimulates preprotein translocation by enhancing the topology inversion of SecG protein.[22] These facts demonstrate that MPIase is widely involved in various membrane integration systems and the translocation of nascent proteins.

2.2.2 Membrane protein integration assay

The membrane protein integration activity was evaluated by the protease protection assay[19,21,23] using in vitro translational system,[24] containing factors for transcription, translation, aminoacylation, and energy regeneration (Fig. 2A). The 3 L-Pf3 coat protein (Fig. 2B)[25] is a mutant of the Pf3 coat protein possessing a longer hydrophobic region due to the addition of three leucine residues. It has been widely used in Sec-independent membrane protein integration research since it does not require membrane potential.[25] We used it as a model substrate protein because it is the simplest one dependent on MPIase.[21] The 3L-Pf3 coat protein possessing ^{35}S-methionine was

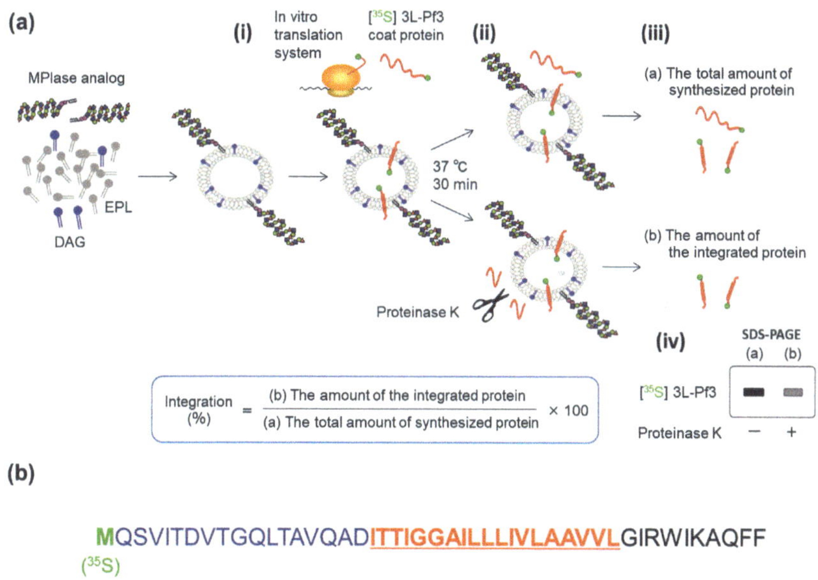

Fig. 2 Membrane protein integration activity assay. (A) (i) Radioisotope-labeled 3L-Pf3 coat proteins are synthesized using an in vitro translation system in the presence of *E. coli* lipids (EPL), liposomes, and the MPIase analog. (ii) The reaction mixture is divided into two fractions, and protease (proteinase K) is added to one. (iii) (a) The protease-free fraction gives a total amount of synthesized protein. (b) The protease-containing fraction gives the integrated protein amount, which is protected from the protease digestion. (iv) The proteins in each fraction are detected via autoradiography after SDS-PAGE. (B) Amino acid sequence of the 3L-Pf3 coat protein. The putative transmembrane region is underlined.

synthesized in the presence of EPL liposomes containing 5 wt% of DAG and 5 wt% of the test compound. When the compound was active, the synthesized protein was integrated into the membrane. After protein synthesis, the reaction mixture was divided into two fractions. One fraction gave the total amount of synthesized proteins via autoradiography of SDS-PAGE. To the other fraction, proteinase K was added to cleave proteins outside the liposomes. Hence, only the integrated protein protected by membranes was detected on SDS-PAGE. The net integration percentage was calculated from (a) the total amount of synthesized protein and (b) the amount of the integrated protein. Normalized integration values were calculated by subtracting the control (EPL/DAG) percentages from the net integration percentages in each experiment.

When we added 5 wt% of the purified MPIase into DAG-containing EPL liposomes, approximately 20% of the 3L-Pf3 coat protein can be integrated into liposomes. The integrated 3L-Pf3 coat protein was increased dose-dependently (0.05–5 wt% of EPL).[21] Moreover, the stoichiometric analysis indicated that a higher number of protein molecules can be integrated than the number of MPIase molecules.[21]

2.2.3 Structural determination of MPIase

We speculated that MPIase contains a proteinaceous moiety because the activity was attenuated by proteinase digestion. However, amino acid analysis of the highly purified MPIase gave no peaks corresponding to amino acids, indicating that MPIase is not proteinaceous (Fig. 3A). Only two peaks were detected, one of which was glucosamine and the other was ammonia. We concluded that some of the amino sugar was decomposed by hydrolysis to give ammonia, since no sugars except glucosamine were detected in the sugar analyses. Both *N*- and *O*-acetyl groups and lipid moieties were detected by ^{1}H and ^{13}C NMR spectroscopy (Fig. 3B).[15] Moreover, two peaks at δ −11.3 ppm and −13.0 ppm in the ^{31}P NMR spectrum suggested the presence of a pyrophosphate diester, which is considered to be a linkage between a glycan chain and a lipid moiety (Fig. 3C). The hydrophobic fraction obtained on 48% aqueous HF hydrolysis of natural MPIase contained the lipid anchor moiety, which was determined to be DAG by Q-TOF-MS (Fig. 3D). Additionally, fatty acids of DAG were similar to that of membrane lipids in *E. coli* (16:0, 16:1, 18:0, and 18:1 etc.) and varied with the culture lots. Thus, we concluded that MPIase is a glycolipid.

The structure of MPIase was determined by combining instrumental analyses and synthetic studies. MALDI-TOF-MS of MPIase produced

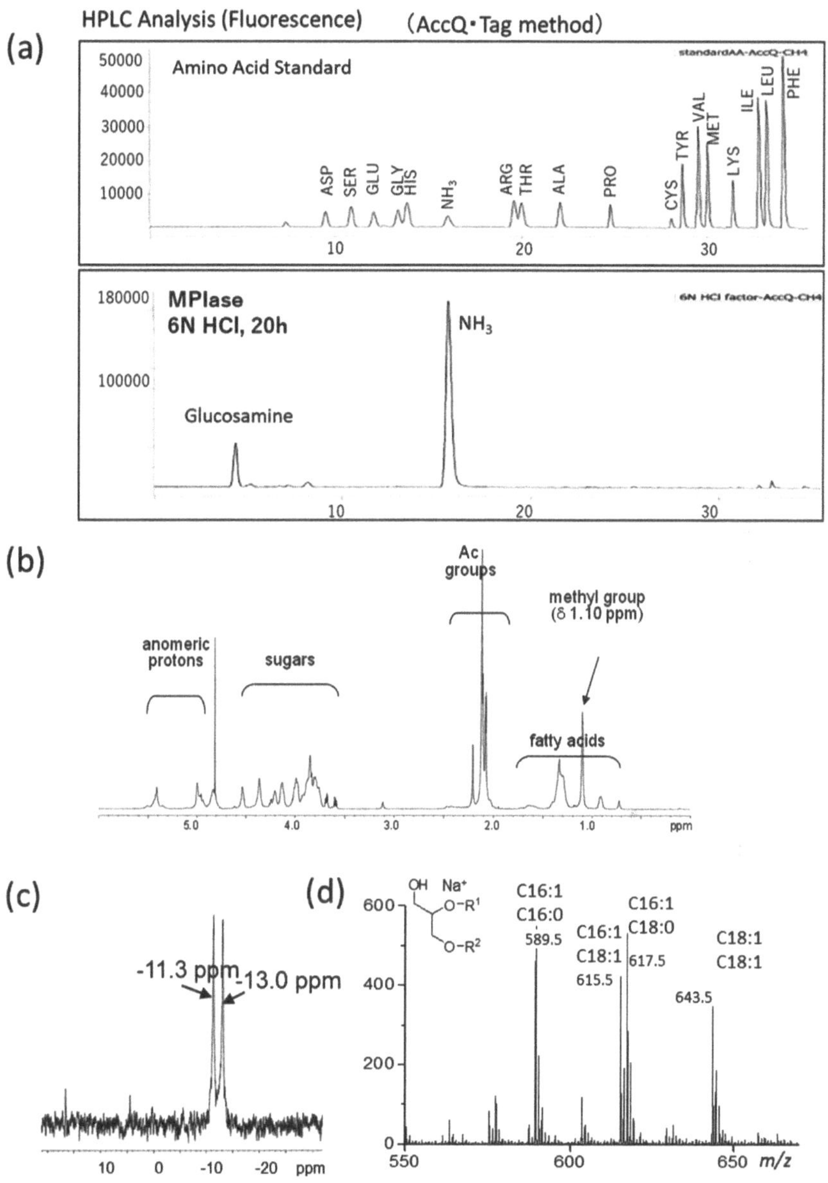

Fig. 3 Determination of the MPlase structure. (A) Amino acid analyses of standard amino acids (top) and MPlase hydrolysate (6 N HCl, 2 h) (bottom). (B) ^{1}H NMR spectrum (750 MHz, D_2O). (C) ^{31}P NMR spectrum (202 MHz, D_2O). (D) Q-TOF-MS of the hydrophobic fraction of MPlase obtained on HF hydrolysis. Diacylglycerols possessing the designated fatty acids were detected as Na^+ adduct ions.

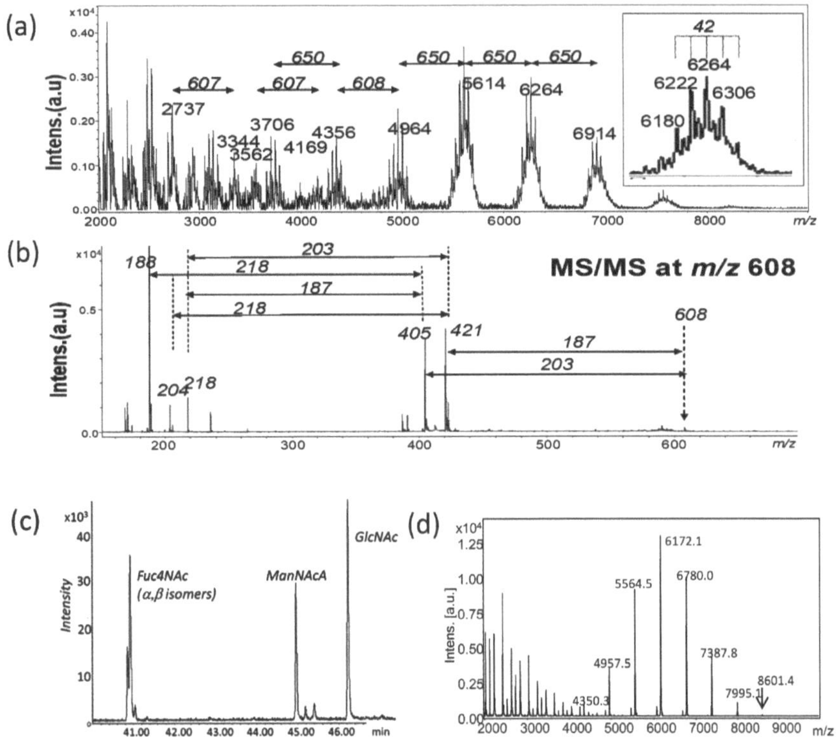

Fig. 4 Structural analyses of MPlase via mass spectrometry. (A) MALDI-TOF-MS spectrum of MPlase in the negative mode. (B) The MS/MS spectrum from *m/z* 608 in the positive mode. (C) GC–MS of MPlase methanolysis products after peracetylation. Three peaks were mainly observed in total ion current chromatography. (D) MALDI-TOF-MS spectrum of MPlase after being hydrolyzed with cold NaOH in the negative mode.

broad peaks, suggesting the diversity of the lipid moiety (Fig. 4A). Moreover, a series of peaks suggested that it contains repeating units, with sizes estimated to be 608 or 650 mass units by the mass intervals between the peak apexes. The subdivided peaks, which varied by 42 mass units, were attributed to the various extents of acetylation in the molecule. MS/MS analysis suggested that the unit possessing *m/z* 608 contains three components with 187, 203, and 217 mass units (Fig. 4B), and that possessing *m/z* 650 contains 187, 245, and 217 mass units. These results may reflect that the 203 mass unit component is partially acetylated. Three peaks appeared in GC–MS after methanolysis supported that MPIase is mainly composed of trisaccharide moieties (Fig. 4C).

The 203 mass unit component corresponds to *N*-acetyl-aminohexose in the dehydrated form. By contrast, the 217 mass unit was presumed to be oxidized, whereas the 187 mass unit would be its deoxy form. Three amide protons were additionally observed by ^{1}H NMR spectroscopy in H_2O, proving that three acetamido sugars are present in each unit. As glucosamine was detected in the amino acid analysis, the 203 mass unit component was identified as *N*-acetylglucosamine (2-acetamido-2-deoxyglucose, GlcNAc). The methyl signal at δ 1.10 correlated with one of the amide protons in NOESY, suggesting a 4-acetamido-4,6-dideoxyhexose. The comparison of NMR spectra and GC–MS results with those of authentic synthetic samples showed that this component was indeed *N*-acetyl-4-amino-4-deoxyfucose (4-acetamido-4,6-dideoxygalactose, Fuc4NAc). The remaining component was assigned as *N*-acetyl-mannosaminuronate (2-acetamido-2-deoxymannuronic acid, ManNAcA) by comparison with some synthetic samples of *N*-acetyl-aminohexuronic acids. Additionally, reduction of the terminal sugar of the glycan gave 2-amino-2-deoxyglucitol, showing the reducing terminal as GlcNAc. All sugars have the D-configuration, as proved on GC–MS analyses of the glycosides obtained on solvolysis in (*S*)-(+)-2-butanol.

The sequence and linkage mode of the trisaccharide were estimated via extensive 2D-NMR measurements and confirmed via syntheses of some substructures. We concluded that the repeating trisaccharide was →(3)-α-D-Fuc4NAc-(1→4)-β-D-ManNAcA-(1→4)-α-D-GlcNAc-(1→). In addition, the NMR spectra of natural MPIase indicated that approximately 30% of GlcNAc is 6-*O*-acetylated.

Lipids and *O*-acetyl moieties were cleaved via mild hydrolysis with cold aqueous NaOH to simplify the structure. The MALDI-TOF-MS of the glycan moiety gave the regular intervals with 608 mass units (Fig. 4D), suggesting that MPIase is a mixture of homologous molecules with the number of repeating units as mainly 9–11. The results mean the glycan of MPIase includes approximately 30 *N*-acetylaminosugars. Taking the aforementioned information into account, we determined the structure of MPIase to be as shown in Fig. 1C.[15]

Notably, the trisaccharide sequence of MPIase is the same as that of the enterobacterial common antigen (ECA) that is present in the outer membranes of enterobacteria.[26] Although the trisaccharide moiety is the same, the glycan of ECA is longer than that of MPIase, and ECA has a monophosphate linkage instead of the pyrophosphate in MPIase. Interestingly, ECA has no membrane protein integration activity despite its similar structure.[15]

3. Syntheses of mini-MPlase-3

To elucidate the action mechanism of MPlase, we synthesized the structurally defined MPlase derivatives since natural MPlase contains heterogeneity in the molecule (the number of OAc groups, the length of the glycan, or types of fatty acids). Our first target compound was the minimal unit of MPlase, a trisaccharyl pyrophospholipid termed mini-MPlase-3 (**1**).[27,28]

3.1 Retrosynthetic strategy

Scheme 1 outlines the design of the synthesis of **1**. The envisaged strategy involved unusual sugars (Fuc4NAc and ManNAcA) syntheses, the construction of a β-mannosaminide bond, selective *O*-acetylation on the 6-OH of GlcNAc, and introduction of a labile pyrophospholipid. Hydrogenation under neutral conditions was considered as the global final deprotection because basic hydrolysis would cleave the 6-OAc groups on GlcNAc, DAG, and pyrophosphate. Thus, benzyl (Bn)-type protecting groups were applied.

Syntheses of the trisaccharide unit of ECA have been reported by Paulsen and Lorentzen[29] and by Boons and coworkers.[30] However, introducing the

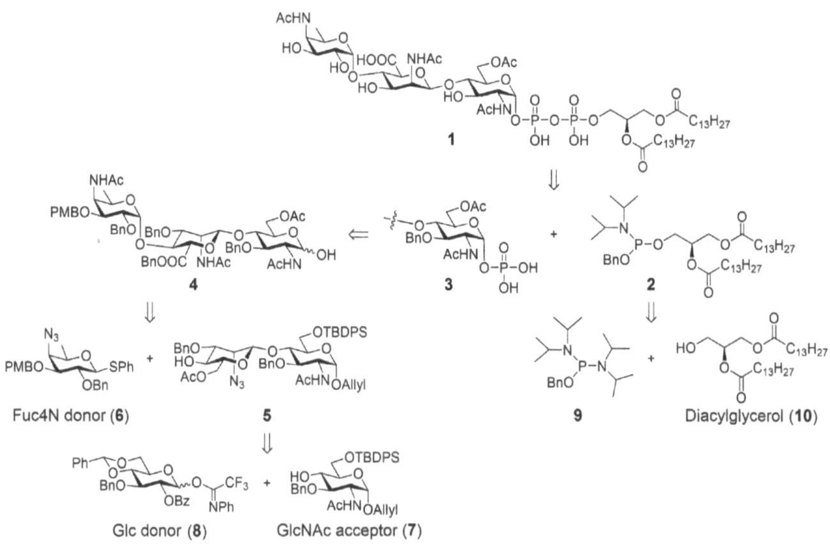

Scheme 1

phospholipid part has not been reported. Since we considered the pyrophospholipid part to play a crucial role in the membrane protein integration process, we planned to introduce the DAG anchor via a labile pyrophosphate near the end of the synthesis. Thus, we designed the phosphoramidite possessing DAG (**2**) and the trisaccharyl phosphate (**3**) as critical intermediates.

Since we had partially synthesized the substructure of the glycan during the structural determination of MPIase,[15] the improved strategy based on that scheme was applied to the synthesis of trisaccharide **4**. The disaccharide acceptor **5** containing β-mannosaminide was constructed by steric inversion at C-2 of the glucose (Glc) residue after obtaining Glc-β-(1→4)GlcNAc. Then, the Fuc4N donor **6** was introduced into the acceptor. The desired trisaccharide structure **4** was obtained by oxidation of the 6-OH of mannosamine to the uronic acid and adding a 6-OAc group on GlcNAc. In the synthesis of the glycan part for ECA, a similar concept was reported by the Boons group.[30]

For fatty acids of DAG, myristic acid (C14:0) was adopted since a saturated fatty acid is convenient for the synthesis, and its phase transition temperature is comparable with that of natural membrane lipids.

3.2 Synthesis of mini-MPIase-3

We started with the synthesis of adequately protected monosaccharide building blocks[31] (Scheme 2). The GlcNAc acceptor **7** had a selectively cleavable 6-OTBDPS group for conversion into an Ac group at the later synthesis stage and a 1-*O*-Allyl group for the selective introduction of the phospholipid moiety. The Glc donor **8**, to be converted into ManNAcA, was synthesized from compound **12**[32] and possessed a 2-OBz group expecting of a neighboring-group participation effect. The Fuc4N donor **6**, whose 3-OH was masked by a selectively removable *p*-methoxybenzyl (PMB) group for future glycan elongation, was prepared from compound **13**[33] via reduction of the 6-OH group and the introduction of an azido group at C-4 in association with the configuration inversion, all in good overall yield.[15]

The glycosylation of the GlcNAc acceptor **7** with the Glc donor **8** selectively produced the desired β-disaccharide **18** (Scheme 3A). Then, removing the 2-OBz group of **18**, introducing the Tf group **19**, and substituting with NaN_3 at C-2 of **19** yielded the disaccharide **20** that contained a mannosamine-type structure in 46% yield over three steps.

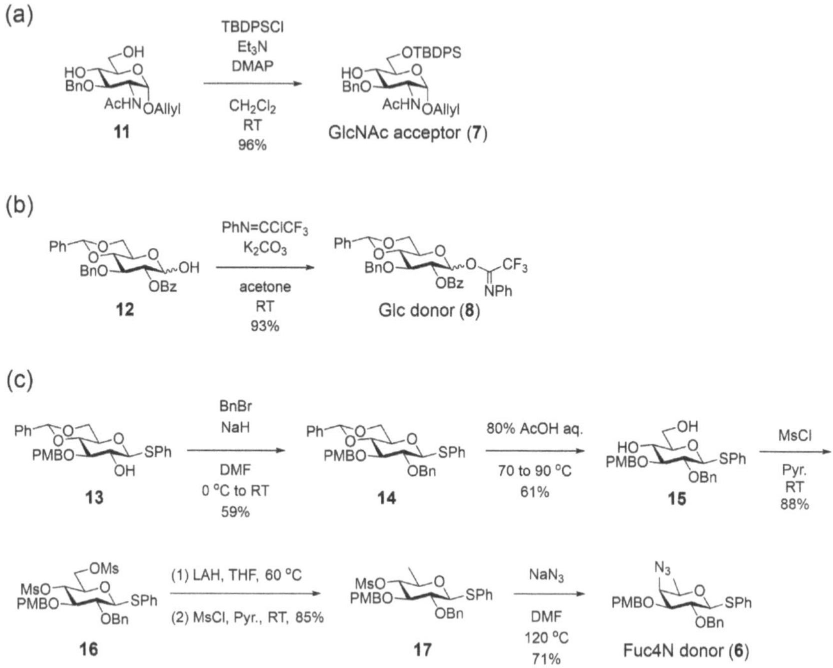

Scheme 2

The deprotection of the benzylidene acetal and regioselective acetylation at the 6-OH of **20** gave the disaccharide acceptor **5**. Then, the condensation of the Fuc4N donor **6** with **5** provided the trisaccharide **21** in 86% yield with good α selectivity (α/β =7/1). Using diethyl ether as a solvent unaltered the stereoselectivity. Removal of the 6-OAc group of the mannosamine moiety **21**, then oxidation of the 6-OH and introduction of the Bn ester afforded the mannosaminuronate **22** in 75% yield over three steps. The TBDPS group of the GlcNAc residue was converted to an Ac group successfully to give the 6-OAcGlcNAc derivative **23**. Although the azido group on the mannose moiety in **23** resisted the Staudinger reaction, that on the fucose moiety was reduced. We also used $NaBH_4$ reduction with $NiCl_2$ or a stepwise Staudinger reaction, where the azido group on the mannose moiety in the disaccharide was first reduced, and then the fucose moiety was reduced after constructing the trisaccharide, in the substructure syntheses. Nevertheless, adopting these procedures might be difficult when the glycan chain is further elongated. Thus, both azido groups of **23** were reduced by combining $SnCl_2$, PhSH, and Et_3N,[34] followed by *N*-acetylation to afford the acetamide as reported by the Boons group.[30] The selective liberation of

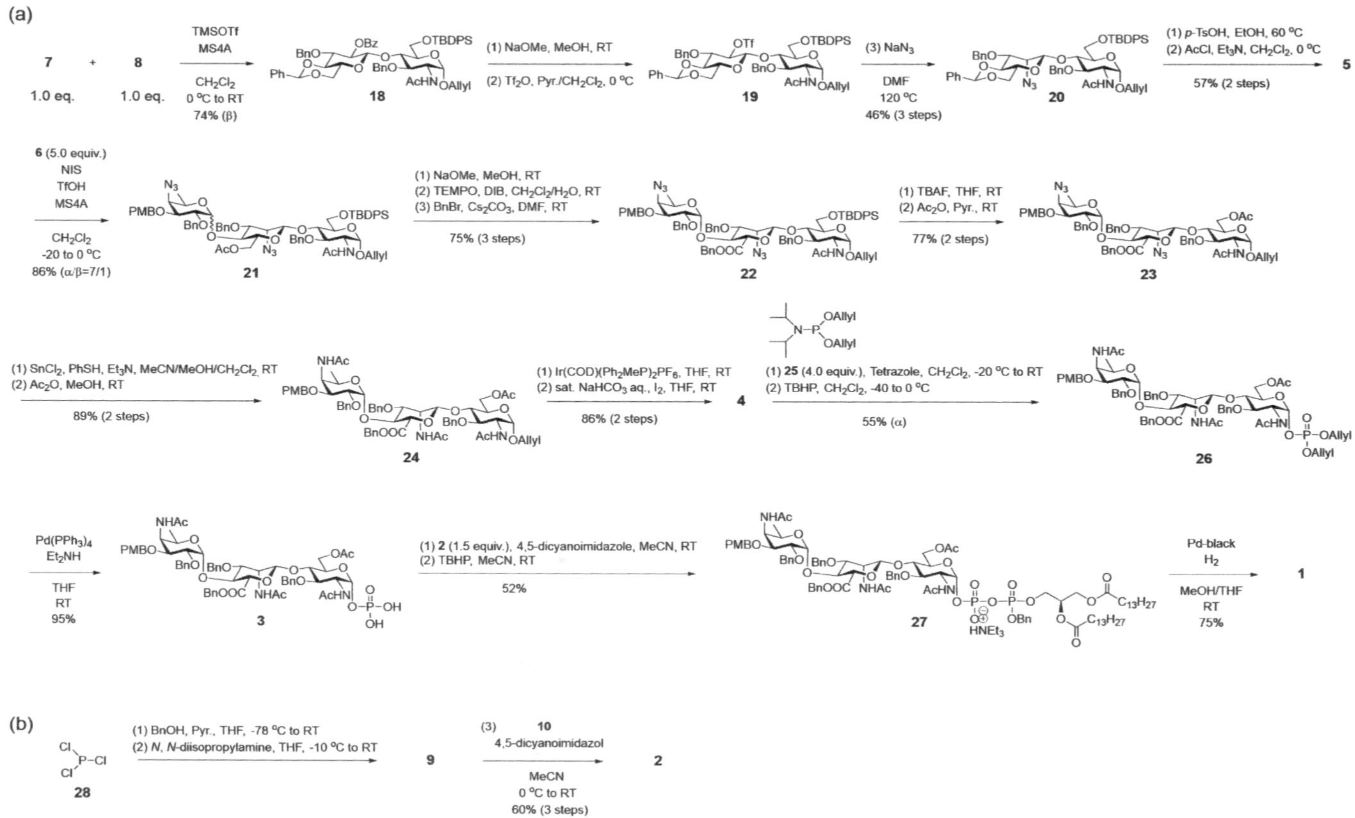

Scheme 3

the hydroxyl group at the reducing end provided the trisaccharide hemiacetal **4**, which was reacted with the commercially available diallylphosphoramidite **25** and was subsequently oxidized with *tert*-butylhydroperoxide produced the phosphoric ester **26**. Only the α-isomer was obtained although **4** was a mixture of α- and β-isomers. Also, the trisaccharide hemiacetal was recovered and the oxazoline was obtained as a byproduct after silica gel purification. Moreover, when the substituent at the 2-position of glucose was an N_3 group instead of the NHAc group, both α- and β-phosphoric esters were isolated, suggesting the decomposition of the β-phosphonate due to participation of the NHAc group during the purification step.[35,36] The allyl groups on the phosphoric acid were removed to obtain trisaccharyl phosphate **3**. In preparing of pyrophospholipids, a morpholine amide of phosphoric acid[37] resulted in a dimerized byproduct. Furthermore, it was difficult to separate it from the desired product because of its high polarity. Thus, we used phosphoramidite **2** having a Bn group and diacylglycerol (Scheme 3B).[38–40] This compound was hydrophobic enough for purification by a silica gel column and could be stored at −20°C for several months.

The fully protected trisaccharylpyrophospholipid **27** was obtained by coupling **3** with **2** in the presence of 4,5-dicyanoimidazole in MeCN at room temperature, followed by oxidation with *tert*-butylhydroperoxide to give the product in 52% yield. Finally, global deprotection of the Bn-type protection groups under hydrogenolysis conditions furnished trisaccharyl pyrophospholipid **1** possessing the 6-OAc group on GlcNAc, namely mini-MPIase-3, in 75% yield.

3.3 Synthesis of mini-MPIase-3 analogs

Trisaccharyl pyrophospholipid without the 6-OAc on GlcNAc, [**29**, mini-MPIase-3(6-OH)] and trisaccharyl DAG [**30**, Trisac-DAG] without a pyrophosphate group were produced similarly (Fig. 5). Additionally, analogs devoid of the DAG anchor [**31**, Trisac-P] and [**32**, Trisac-P(6-OH)] were synthesized from the common intermediates. Furthermore, we prepared some derivatives from natural MPIase. The glycan moiety of natural MPIase attached with a monophosphate group [**33**, Polysac-P] was obtained by pyrophosphatase digestion. By contrast, the polysaccharide moiety lacking a phosphate group [**34**, Polysac] was gained through the hydrolysis of natural MPIase with AcOH.

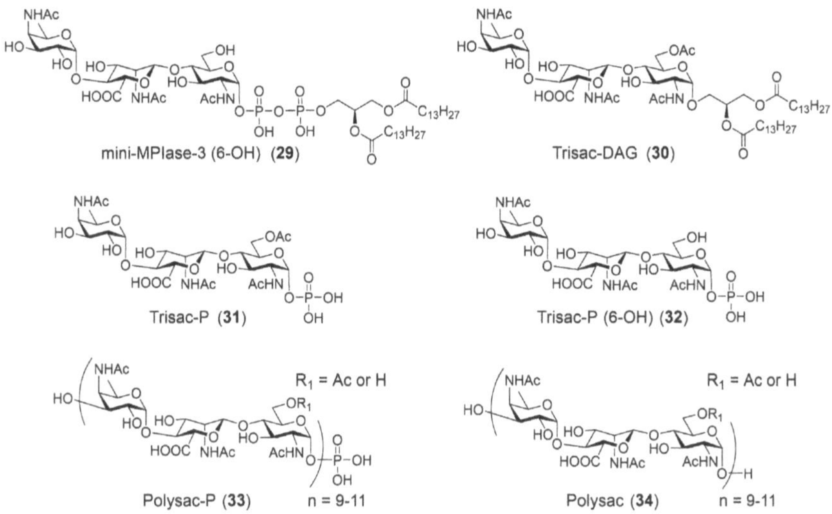

Fig. 5 Structures of MPlase derivatives.

4. Structure–activity relationship studies

4.1 Membrane protein integration by mini-MPlase-3

As mentioned above, the membrane protein integration assay demonstrated that approximately 20% of the 3L-Pf3 coat protein integrated into the liposomes with the assistance of natural MPIase. Although the synthetic analog **1** was weaker than natural MPIase, significant activity ($5.2 \pm 0.8\%$) indicated that it contains the minimum active structure (Fig. 6).[27] Notably, the glycan of **1** is only approximately 1/10th of that of the natural MPIase, whereas **29** and **30** were less potent ($2.0 \pm 0.5\%$ and $2.4 \pm 0.6\%$, respectively). These results demonstrated that glycan length, the 6-OAc group on GlcNAc, and the phosphate group all contribute to the membrane protein integration.

We also observed that other Sec-independent substrate proteins, such as wild-type Pf3 phage coat protein and M13H5 phage procoat protein,[10] integrated into the liposomes by natural MPIase or **1**, but not by **29**.

4.2 Chaperone-like activity of the glycan part of MPlase

Next, the role of the lipid anchor moiety in MPIase was investigated. When MPIase derivatives lacking the lipid moiety (**31**, **32**, **33**, **34**) were added to

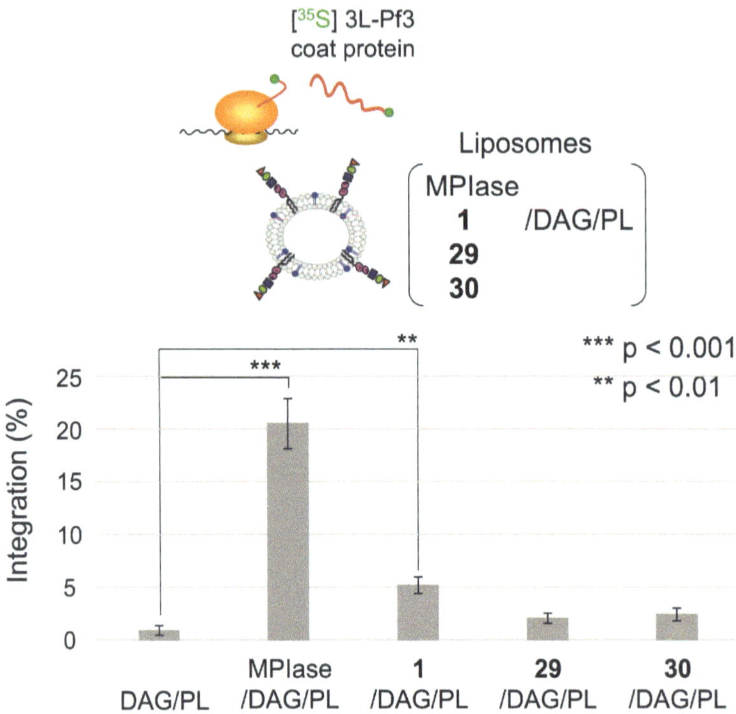

Fig. 6 Integration activity of MPlase derivatives. Net percentages of proteins integrated into liposomes composed of DAG/EPL, MPlase/DAG/EPL, **1**/DAG/EPL, and **29**/DAG/EPL.

the reaction mixture containing the control liposomes (DAG/EPL), none showed integration activity even at high concentrations of anchorless derivatives (1.0 mg/mL) (Fig. 7A).[27] This suggested that MPIase must be affixed in the membrane.

On the other hand, **33** significantly enhanced the integration when the liposomes contained 5% of **1**, whereas the membrane protein integration activity of **1** decreased by the addition of **31**. These results supported the idea that the glycan interacts with the protein in solution. However, these synergistic effects were not observed by adding of other derivatives without the phosphate or 6-OAc group, demonstrating that these groups are important for the interaction. Hence, we assumed that **33** prevented the aggregation of proteins and increased their integration-competent structures, whereas **31** was too short to cover the hydrophobic region of the protein.

Our hypothesis was also verified by gel-filtration experiments (Fig. 7B).[15] In the absence of **33**, the hydrophobic 3L-Pf3 coat protein synthesized by the cell-free protein translation system was immediately aggregated and eluted in

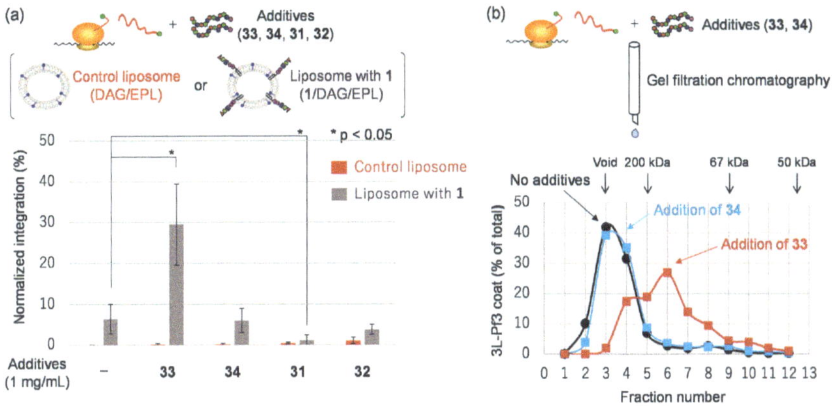

Fig. 7 Integration activity of MPIase derivatives. (A) Normalized integration values by adding anchorless MPIase derivatives into a reaction mixture with liposomes containing **1**/DAG/EPL. (B) The amount of 3L-Pf3 coat protein in each fraction after gel-filtration chromatography is plotted as a percentage of the total amount.

the void fraction. By contrast, the protein was eluted in the lower molecular weight fractions in the presence of **33**, indicating that **33** formed a soluble complex with the 3L-Pf3 coat protein and prevented aggregation (chaperone-like activity[41]). Such an anti-aggregation effect was not found for **34**. Therefore, the phosphorylated long glycan chain is necessary for the anti-aggregation effect.

5. Intermolecular interactions between MPIase and the substrate proteins

5.1 Inhibition of protein aggregation by MPIase

Sec-independent proteins generated from ribosomes immediately aggregate if there is no assistance. We assumed that MPIase should prevent proteins from aggregating like a chaperone because ribosome-associating chaperones have not been identified in *E. coli*. Thus, we verified the direct interaction between MPIase and the substrate protein.[42] We used the Pf3 coat protein[10] (Fig. 8A) without the 3L-substituent since extreme hydrophobicity of 3L-Pf3 coat protein hampered the purification. When we prepared the Pf3 coat protein solutions, even a dilute solution caused fluffy insoluble matter to form (Fig. 8B). However, the Pf3 coat protein dispersion in an MPIase-containing buffer generated no visible, insoluble matter. Since MPIase inhibited forming aggregates of the Pf3 coat proteins, we measured the CD spectra of the proteins to detect the alteration of the secondary

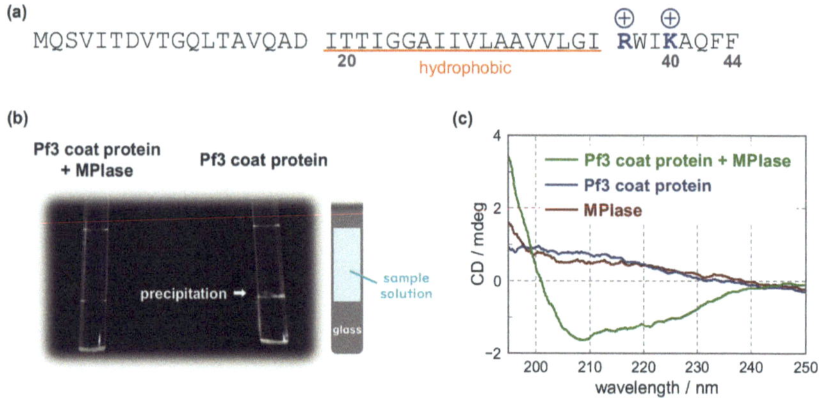

Fig. 8 Inhibition of protein aggregation by MPIase. (A) The amino acid sequence of the Pf3 coat protein. (B) The Pf3 coat protein solutions in a phosphate buffer with (left) or without (right) MPIase. (C) CD spectra of the Pf3 coat proteins in a phosphate buffer with or without MPIase.

structure. The CD spectra showed significant difference in the presence and absence of MPIase (Fig. 8C). The Pf3 coat protein in the MPIase-free buffer did not form definite secondary structures, whereas the characteristic of an α-helix is shown in the MPIase-containing buffer. These results revealed that MPIase alters the secondary structure of a substrate protein, deterring it from becoming a random aggregate structure, consequently, promoting its integration into lipid membranes.

5.2 Analysis of the interaction between MPIase and the Pf3 coat protein

5.2.1 Analyses of the intermolecular interactions by SPR

We evaluated the interaction between MPIase and the Pf3 coat protein in the aqueous solution using surface plasmon resonance (SPR). First, the Pf3 coat proteins were immobilized on the flow cells of the sensor chip and solutions of MPIase analogs flowed over the chip as analytes.[43] The direct interaction between MPIase and proteins was confirmed since natural MPIase attached to the Pf3 coat protein in a concentration-dependent manner (Fig. 9A). Dissociation constants (K_D) were determined using a steady-state affinity model (2.6 ± 0.6 μM).

Next, we compared the synthetic analogs, **1** and **30** (Fig. 9B). The results showed that both analogs interacted with the Pf3 coat protein in a manner dependent on the concentration; however, the response amplitude of **30** was weaker than that of **1**. Moreover, the K_D value of **30** (136.7 ± 32.7 μM) was

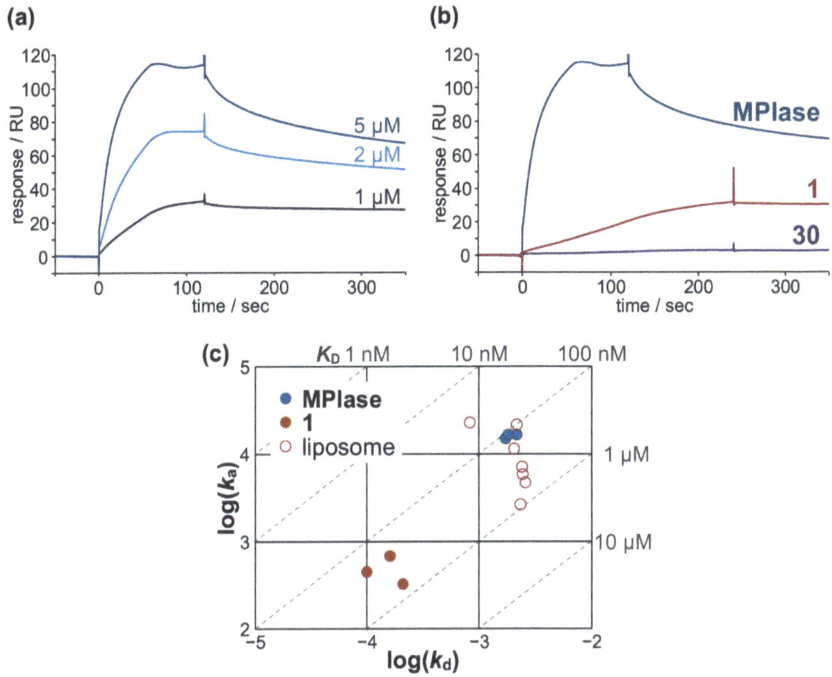

Fig. 9 SPR analyses to evaluate the interaction between the Pf3 coat proteins and MPlase analogs. (A) Sensorgrams showing the binding of natural MPlase to the Pf3 coat protein. (B) Comparisons among sensorgrams corresponding to 5 μM of MPlase, **1**, and **30**. (C) Kinetic profiles of MPlase and **1** presented in an on–off rate map. The apparent association rates (k_a) are plotted against the apparent dissociation rates (k_d) using double logarithmic coordinates. The diagonal lines represent equal dissociation constants (K_D). The values of **1** in liposomes were also plotted as open circles.

larger than that of **1** (2.1 ± 1.6 μM), signifying the importance of a pyrophosphate group of MPlase.

Although the K_D values of MPlase and **1** were comparable in SPR, MPlase had higher activity than **1** in the membrane integration assay. Analysis of their kinetics of binding revealed the difference between these two glycolipids. The association and dissociation rates were different as shown in the on–off rate map (Fig. 9C), which visually characterizes the binding properties.[44] Both the association and dissociation of MPlase to the Pf3 coat protein were more rapid than those of **1**. A long sugar chain can contribute to rapid association and dissociation.

Additionally, we studied the interactions of EPL liposomes containing **1**. Although the lipid moiety of **1** was masked in the membranes, it bound to the protein. The apparent association and dissociation rates of **1** contained in liposomes were faster than those in the solution. A high local concentration of **1** in the membranes might hasten the interactions by assembling the trisaccharides like a long glycan chain of MPIase. Although we used the 1:1 binding model in this study, notably, SPR cannot precisely determine the affinity and kinetic parameters when the stoichiometry of the molecular interaction is unknown. Nonetheless, our study showed that the apparent equilibrium rate depends on the chain length or the local concentrations.

5.2.2 Interactions of the phosphorylated trisaccharides

In addition, saturation-transfer difference (STD)-NMR spectroscopy[45] and docking simulations were performed to precisely investigate the contribution of partial structures in the phosphorylated glycan on the intermolecular interaction with a protein.

Trisaccharide analogs without a lipid moiety were used for STD-NMR experiments. First, STD signals were detected from most protons of the trisaccharide (Fig. 10A), indicating that the Pf3 coat protein and trisaccharide can bind in various modes adopting possible conformations. Next, we compared the relative intensities of the STD signals, where the fractional saturation was normalized with that of the methyl group. Although the relative intensities of the 6-*O*-acetyl and *N*-acetyl groups on the GlcNAc unit of **31** were significantly greater than those of the acetyl groups on Fuc4NAc/ManNAcA and the methyl group on Fuc4NAc, the values of **32** were comparable (Fig. 10B).

Docking simulations of the virtual ligands (**35, 36,** and **37**) were then performed on each residue of the Pf3 coat protein. The cumulative frequencies of the contact residues to each ligand in docking poses with higher docking scores showed that the Pf3 coat protein interacted better with **35** than with **36**, and hardly contacted with **37** (Fig. 10C). Heat map expanding the cumulative frequencies indicated the preferred combinations. The importance of the 6-*O*-acetyl group on GlcNAc and pyrophosphate for the interaction was notable because **35** frequently contacted with residues in the hydrophobic region and with the basic amino acid residues, Arg37 and Lys40, at the C-terminus of the protein.

5.2.3 Verification of the interacting factors

Since docking simulations illustrated the contacts between the trisaccharide and hydrophobic region and the basic amino acids of the Pf3 coat protein,

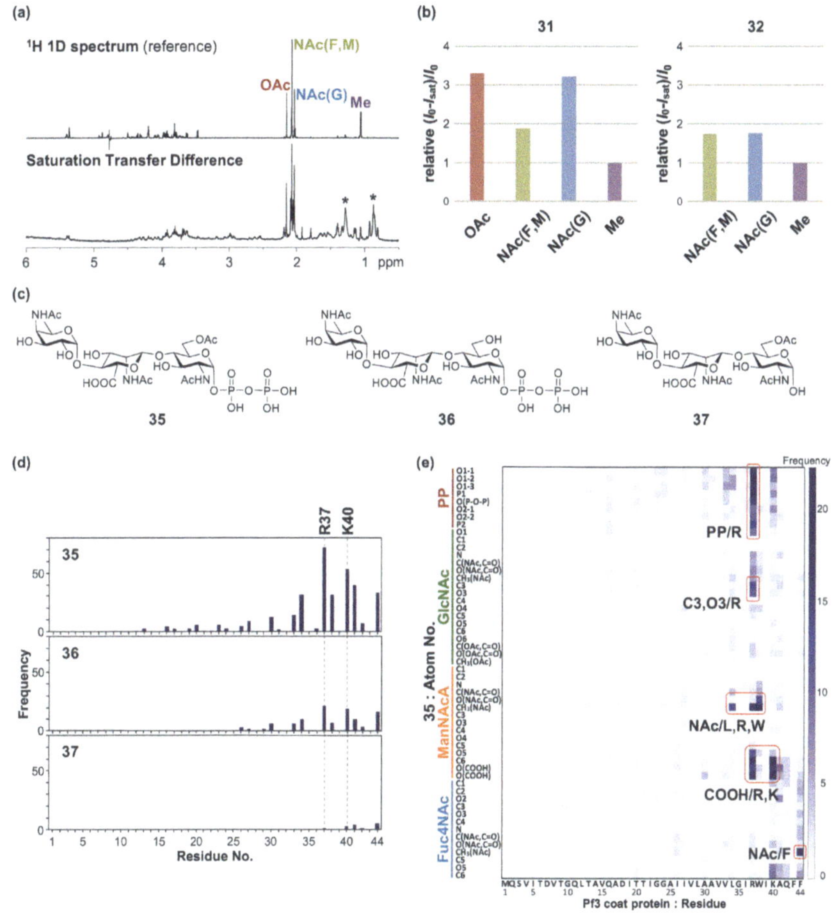

Fig. 10 Interaction analyses of trisaccharide analogs with the Pf3 coat protein. (A) STD-NMR spectrum represents the interaction between trisaccharide with phosphate **31** and the Pf3 coat protein (bottom). A one-dimensional ^{1}H NMR spectrum of **31** is shown for comparison (top). Signals of acetyl and methyl groups were assigned to the structure of **31**: OAc, 6-*O*-acetyl group on GlcNAc; NAc(F,M), *N*-acetyl groups on Fuc4NAc and ManNAcA; NAc(G), *N*-acetyl group on GlcNAc; Me, methyl group on Fc4NAc. The *N*-acetyl groups on Fuc4NAc and ManNAcA overlapped with each other. (B) Normalized intensities of STD signals from STD-NMR analyses of **31** and **32** with the Pf3 coat proteins. (C) Structures of trisaccharide analogs for the docking simulation. (D) The cumulative frequencies of the Pf3 coat protein residues found at the interfaces of **35**, **36**, and **37** were obtained from computational docking simulations. (E) Heat map of frequencies in the docking simulation is shown over the atoms of **35** (vertical axis) and the amino acid sequence of the protein (horizontal axis).

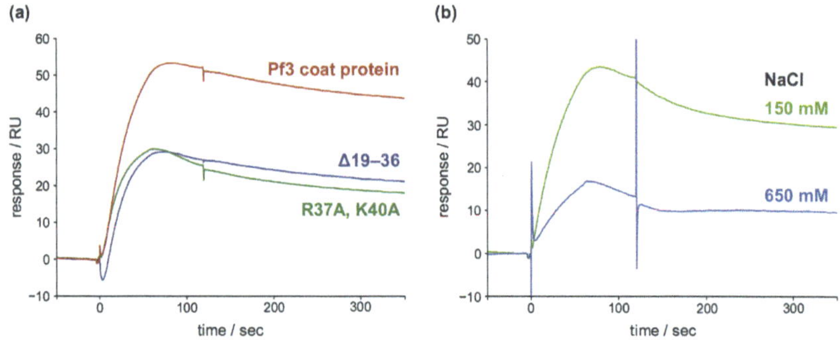

Fig. 11 SPR analyses to evaluate the interacting factors. (A) SPR sensorgrams showing the binding of natural MPIase to three Pf3 coat protein mutants, the Pf3 coat protein, Pf3(Δ19–36), and Pf3(R37A, K40A). (B) SPR sensorgrams showing the binding of natural MPIase to the Pf3 coat protein in a conventional buffer or a high-salt concentration buffer.

we measured SPR using the Pf3 coat protein mutants without each region. Consequently, the binding of MPIase decreased compared with that of the intact Pf3 coat protein (Fig. 11A). Thus, it was confirmed that MPIase captures the hydrophobic region and the basic amino acids in the Pf3 coat proteins. Additionally, the interaction between the Pf3 coat protein and MPIase in a high-salt buffer (650 mM NaCl) was lesser than that under the conventional conditions (150 mM) (Fig. 11B). The electrostatic interactions involving the pyrophosphate and basic amino acid residues would be disturbed in the high-salt solution.

6. Membrane protein integration mechanism involving MPIase

We proposed the mechanism of activity on the basis of the structure–activity relationship studies (Fig. 12).[15,27] Without MPIase, a hydrophobic protein released from a ribosome rapidly aggregates and fails to integrate. By contrast, with MPIase, the glycan moiety of MPIase captures a protein, thus, preventing aggregation on the membrane surface. Also, the hydrophobic interactions between the acetyl groups of MPIase and hydrophobic regions of the protein, combined with the electrostatic interactions between the pyrophosphate of MPIase and the basic residues of the protein are vital for promoting membrane protein integration. Furthermore, firm negative charges of the pyrophosphate linkage would attract the positive charges in the protein.[46–48] Generally, membrane proteins operate with

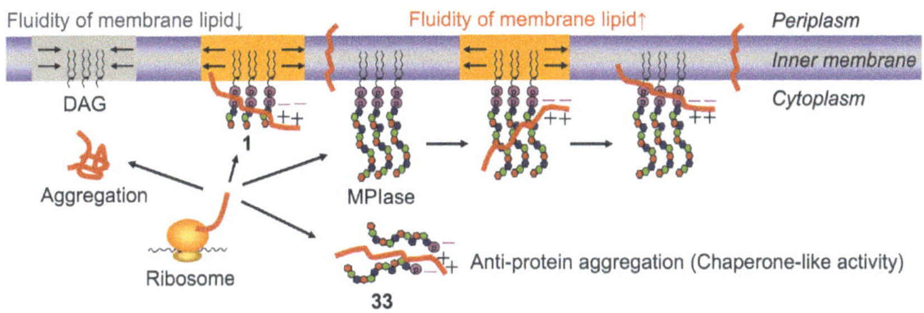

Fig. 12 Initial step of the translocon-independent membrane protein integration model. The endogenous DAG blocks spontaneous integration by attenuating the membrane mobility and the substrate protein aggregates. In the presence of MPIase, the mobility of the membrane is increased. The glycan chain of MPIase binds to the protein by using the hydrophobic and electrostatic interactions, and thus, the protein aggregation is inhibited. The substantial negative charges of the pyrophosphate group attract the protein possessing positive charges to the membrane surface. Since **1** captures the protein near the membrane surface, the protein can proceed into the loosened membrane.

the "positive-inside rule," where cytoplasmic loops near TM domains possess basic amino acid residues.[49] Thus, the interactions are retained in most substrate proteins. More potent hydrophobic interactions between membrane lipids and the hydrophobic part of proteins on the membrane surface cause the integration.

Recently, we observed that DAG and MPIase alter the physicochemical properties of membranes.[50,51] DAG makes the membrane more highly ordered, and its rapid flip-flop motion reduces the exposure of hydrophobic regions in the membrane. On the other hand, when MPIase is present together with DAG, the flexible glycan of MPIase considerably loosened the membrane more, exposing the defects in the membrane. Thus, MPIase enables proteins to contact the inside of the membrane.[52]

Although a single trisaccharide unit in **31** cannot prevent aggregation, **1** has a weak ability in the integration activity. Since **1** captures an unaggregated protein near the membrane surface, it can quickly transfer it into the membrane before aggregation.

7. Biochemical analyses of MPIase functions

7.1 Biosynthesis of MPIase

Despite in vitro evaluation of the function of MPIase, the lack of information on its biosynthesis hampered verification of its involvement in vivo.

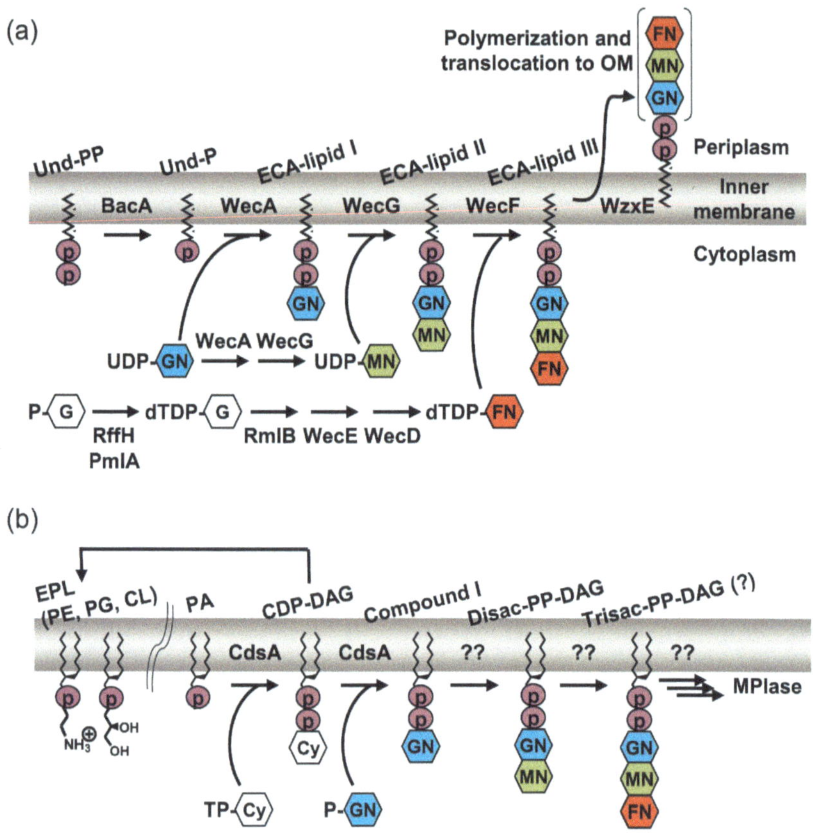

Fig. 13 (A) Biosynthetic pathway of ECA. (B) Putative biosynthetic pathway of MPIase. P: Phosphate, G: Glc, GN: GlcNAc, MN: ManNAc, FN: Fuc4NAc, Cy: Cytidine, Und: Undecaprenol.

Although ECA has the same trisaccharide unit as MPIase, the knockout of each biosynthetic gene for ECA did not affected MPIase expression, showing the presence of a different gene set for MPIase biosynthesis.[53] The glycan unit for ECA is biosynthesized from undecaprenol and sugar-nucleotides[54] (Fig. 13A). Reportedly, depletion of Fuc4NAc accumulated ECA-lipid II (Fig. 13A) as well as disac-PP-DAG[55] (Fig. 13B), suggesting that DAG is used as a carrier lipid for MPIase instead of undecaprenol. Thus, we assumed that the first step of MPIase biosynthesis was the generation of compound I from phosphatidic acid (PA)[56] (Fig. 13B). Therefore, we identified the enzyme that catalyzes the conversion of PA to compound I. We found that CdsA, a CDP-DAG synthase responsible for phospholipid biosynthesis,[57,58] and its paralog YnbB are involved in the generation of compound I.[59] Depletion of CdsA resulted in the MPIase deletion, protein integration and preprotein translocation disorders, and *E. coli* cell-growth inhibition even if a mitochondrial enzyme, Tam41p,[60] compensated for CDP-DAG biosynthesis. These results demonstrated that MPIase is also essential for growth and is involved in protein translocation and membrane integration in vivo. Under cold conditions (<25°C), the expression level of MPIase increased for the efficient protein translocation across membranes, caused by the upregulation of CdsA and YnbB.[61,62] Protein translocation and integration are intrinsically cold-sensitive.[63] It is explained that the reason is that membrane fluidity becomes lower in the cold. However, the expression level of the integration factors except MPIase remains unaffected upon the temperature downshift.[61,62,64] Hence, it is possible that MPIase suppresses the cold sensitivity of protein translocation and integration by increasing its expression level.

7.2 Cooperative activity with YidC

YidC was identified as a protein that interacts with the Sec translocon.[65] Its homologs, Oxa1p in mitochondria and Alb3p in chloroplasts, are involved in membrane protein integration.[66] Since YidC is also involved in Sec-independent integration in *E. coli*, it was proposed as a membrane protein insertase.[67,68] However, no integration activity by YidC has been observed in our translational assay in vitro using DAG-containing proteoliposomes without MPIase. On the other hand, MPIase promoted the membrane protein integration without YidC.[19,20] Meanwhile, YidC accelerated the integration of membrane proteins in combination with MPIase,[69–71] especially when the substrate protein is increased. We considered that MPIase functions at an initial step of protein integration into the membrane, while YidC allows complete insertion at a later step (Fig. 14A). This mechanism suggests a functional interaction between MPIase and YidC. Consistently, MPIase is upregulated upon YidC depletion,[70] strongly suggesting that MPIase partially suppressed the defect caused by YidC depletion. Pull-down analysis revealed that MPIase is co-precipitated with YidC, indicating a direct interaction between MPIase and YidC.[72]

The single-crystal X-ray structures show that the TM helices of YidC form a hydrophilic groove that is open toward both the lipid bilayer and the cytoplasm.[73–75] YidC would receive a part of the substrate through electrostatic interaction between the positive charge inside the cavity of YidC

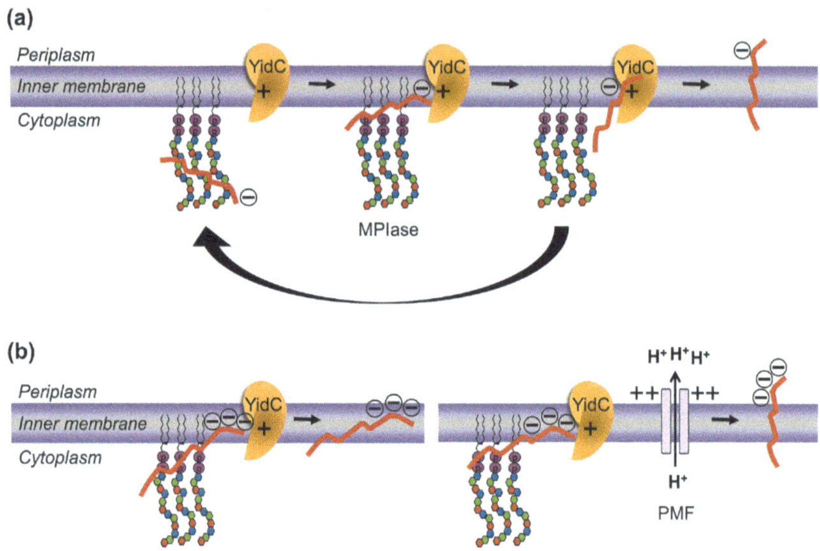

Fig. 14 Later steps of the translocon-independent membrane protein integration model in *E. coli*. (A) Cooperative action of MPlase and YidC. YidC receives a substrate protein from MPlase through electrostatic interaction between the positive charge inside the cavity of YidC and the negative charges of the protein. As the protein dissociates from MPlase, MPlase can be ready to accept the next substrate. (B) Assistance of proton motive force (PMF). (left) When the negative charges in the N-terminal regions increase or the expression level of the substrate increases, only MPlase is insufficient for the complete insertion. (right) PMF accelerates the step in which the N-terminal region is translocated to the periplasmic space to complete insertion. PMF also powers the YidC function.

and the negative charges of substrate proteins, which would accelerate the dissociation of the protein from MPIase to complete the integration. Additionally, proton motive force (PMF) stimulated the MPIase-dependent integration if the substrate synthesis level increased or negative charges in the N-terminal region of N-out topology substrate protein increased (Fig. 14B).[72] These results suggested that MPIase, YidC, and PMF function sequentially and cooperatively during the catalytic cycle of membrane protein integration.

7.3 Cooperative activity with SecYEG

MPIase also functions cooperatively with SecYEG. The SecA-dependent preprotein translocation is expressively stimulated when MPIase is co-reconstituted with SecYEG.[19,22] Considering that MPIase drives the membrane insertion of Sec-independent substrates, it is plausible that

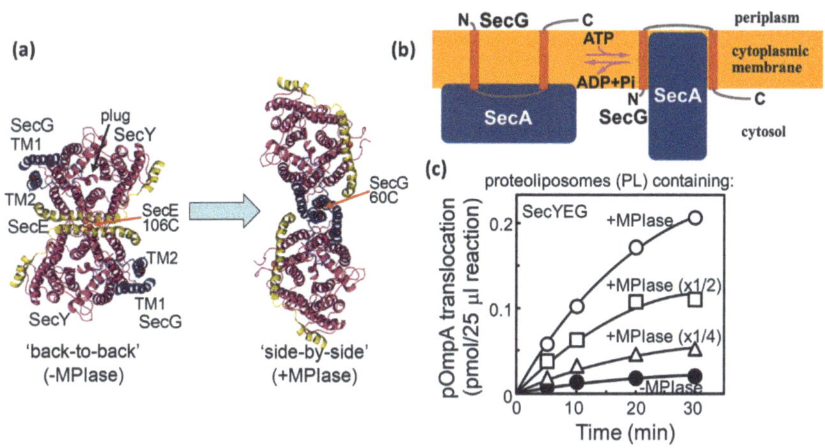

Fig. 15 Interplay between MPIase and SecYEG. (A) The SecYEG dimer models. The structure of the SecYEG protomer was modeled into back-to-back (left) and side-by-side (right) dimers. MPIase altered the orientation of the SecYEG dimer. The SecE 106C mutant forms a homodimer in the absence of MPIase upon oxidization whereas the SecG 60C mutant does in the presence of MPIase. SecY, purple; SecE, yellow; SecG, blue. (B) Topological arrangement of SecG. In the absence of preprotein translocation, both the N- and C-terminal regions of SecG are exposed to the periplasmic space. By contrast, during SecA-dependent translocation, SecG undergoes a repeat of the topology inversion cycle to accelerate the SecA-dependent translocation. MPIase is essential for SecG inversion. (C) Translocation of pOmpA into proteoliposomes. Translocation of a presecretory protein, pOmpA, into the proteoliposomes reconstituted with SecYEG is stimulated by MPIase.

MPIase may facilitate membrane insertion of the signal peptide of preproteins. In addition, we found that MPIase modulates the structure of the SecYEG dimer by directly affecting SecYEG[22] (Fig. 15A). It is also shown that MPIase is co-precipitated with SecYEG. SecYEG forms a dimer in which SecE lies at the dimer interface in the absence of MPIase (the "back-to-back" structure). By contrast, MPIase transforms SecYEG into the structure where SecG lies near the dimer interface (the "side-by-side" structure). Besides, SecG stimulates preprotein translocation by undergoing the cycle of topology inversion, which causes the structure change of SecA smoother (Fig. 15B).[22,76] The SecG inversion only occurs when SecYEG forms the side-by-side structure in the presence of MPIase. Summarily, MPIase transforms SecYEG into the side-by-side structure, with which the topology inversion cycle of SecG operates, resulting in the considerable stimulation of preprotein translocation (Fig. 15C).

MPIase is involved in the Sec-dependent insertion as well.[19,70] However, it is unclear whether or not the side-by-side structure is also necessary for membrane insertion. Considering that all the translocons do not form a dimer, it seems not the case. Alternatively, it is possible that MPIase functions as a membrane protein integrase for Sec-dependent substrates, as well as the case for the Sec-independent insertion. SecYEG would serve as a ribosome receptor. With the ribosome associating with SecYEG, the growing nascent chain of membrane proteins would insert into the membrane with the aid of MPIase, and subsequently of YidC.

8. Conclusions

MPIase is the first known glycolipid involved in protein translocation and membrane protein integration. It has characteristic constituent sugars, many *N*- and *O*-acetyl groups, pyrophosphoric acid, and diacylglycerol. The combination of these components seems to function on the translocon-independent integration. MPIase is also an essential factor for the translocon-dependent integration and translocation pathways. Thus, MPIase would interact with various substrate proteins and cooperate with Sec factors to drive preprotein translocation and membrane protein integration. Depletion of MPIase biosynthesis demonstrated its importance in cell viability. However, localization of MPIase in *E. coli* remains unclear because the expression level of MPIase is low (isolation yield is 0.5 wt% of EPL; i.e., <1/1000 of membrane lipids based on the number of molecules). Therefore, the assembly of MPIase molecules on the membrane and co-localization with other factors should be verified using a combination of synthetic chemistry, biochemistry, physicochemical measurements, and molecular-biological approaches.

We shed light on the new biological functions of glycolipids. Elucidating the molecular basis of protein translocation and membrane protein integration in *E. coli* would improve in the development of antibacterial agents, protein aggregation inhibitors, and membrane-protein reconstruction techniques.

Acknowledgments

We thank all of the collaborators who were involved in the research described herein. We dedicate this article to the memory of Professor Hidetoshi Yamada (Kwansei Gaskin University) for his significant contributions to the field.

References

1. Blobel, G. Protein Targeting (Nobel Lecture). *Chembiochem* **2000**, *1*, 86–102.
2. Rapoport, T. A.; Goder, V.; Heinrich, S. U.; Matlack, K. E. Membrane–Protein Integration and the Role of the Translocation Channel. *Trends Cell Biol.* **2004**, *14*, 568–575.
3. Koch, H. G.; Moser, M.; Muller, M. Signal Recognition Particle-Dependent Protein Targeting, Universal to All Kingdoms of Life. *Rev. Physiol. Biochem. Pharmacol.* **2003**, *146*, 55–94.
4. Chen, Y.; Shanmugam, S. K.; Dalbey, R. E. The Principles of Protein Targeting and Transport Across Cell Membranes. *Protein J.* **2019**, *38*, 236–248.
5. Rapoport, T. A. Protein Translocation Across the Eukaryotic Endoplasmic Reticulum and Bacteria Plasma Membranes. *Nature* **2007**, *450*, 663–669.
6. Kuhn, A.; Koch, H.-G.; Dalbey, R. E. Targeting and Insertion of Membrane Proteins. *EcoSal Plus* **2017**, 7 (2). https://doi.org/10.1128/ecosalplus.ESP/0012-2016.
7. Serek, J. *Escherichia coli* YidC Is a Membrane Insertase for Sec Independent Proteins. *EMBO J.* **2004**, *23*, 294–301.
8. Kuhn, A.; Kiefer, D. Membrane Protein Insertase YidC in Bacteria and Archaea. *Mol. Microbiol.* **2017**, *103*, 590–594.
9. Kuhn, A.; Haase, M.; Leptihn, S. Assisted and Unassisted Protein Insertion Into Liposomes. *Biophys. J.* **2017**, *113*, 1187–1193.
10. Kuhn, A. Major Coat Proteins of Bacteriophage Pf3 and M13 as Model Systems for Sec-Independent Protein Transport. *FEMS Microbiol. Rev.* **1995**, *17*, 185–190.
11. Hegde, R. S.; Keenan, R. J. Tail-Anchored Membrane Protein Insertion Into the Endoplasmic Reticulum. *Nat. Rev. Mol. Cell Biol.* **2011**, *12*, 787–798.
12. Borgese, N.; Coy-Vergara, J.; Colombo, S. F.; Schwappach, B. The Ways of Tails: The GET Pathway and More. *Protein J.* **2019**, *38*, 289–305.
13. Peschke, M.; Le Goff, M.; Koningstein, G. M.; Vischer, N. O.; Abdel-Rehim, A.; High, S.; van Ulsen, P.; Luirink, J. Distinct Requirements for Tail-Anchored Membrane Protein Biogenesis in *Escherichia coli*. *MBio* **2019**, *10*, e01580. https://doi.org/10.1128/mBio.01580-19.
14. Geller, B. L.; Wickner, W. M13 Procoat Inserts Into Liposomes in the Absence of Other Membrane Proteins. *J. Biol. Chem.* **1985**, *260*, 13281–13285.
15. Nishiyama, K.; Maeda, M.; Yanagisawa, K.; Nagase, R.; Komura, H.; Iwashita, T.; Yamagaki, T.; Kusumoto, S.; Tokuda, H.; Shimamoto, K. MPIase Is a Glycolipozyme Essential for Membrane Protein Integration. *Nat. Commun.* **2012**, (*3*), 1260. https://doi.org/10.1038/ncomms2267.
16. Nishiyama, K.; Shimamoto, K. Glycoli Lipozyme Protein Integrase (MPIase): Recent Data. *Biomol. Concepts* **2014**, *5*, 429–438.
17. Shimamoto, K. Elucidation of Excitatory Neurotransmission and Membrane Protein Integration Mechanisms. *Bull. Chem. Soc. Jpn.* **2016**, *89*, 282–295.
18. Koch, H.-G.; Hengelage, T.; Neumann-Haefelin, C.; MacFarlane, J.; Hoffschulte, H. K.; Schimz, K.-L.; Mechler, B.; Muller, M. *In Vitro* Studies With Purified Components Reveal Signal Recognition Particle (SRP) and SecA/SecB as Constituents of Two Independent Protein-Targeting Pathways of *Escherichia coli*. *Mol. Biol. Cell* **1999**, *10*, 2163–2173.
19. Nishiyama, K.; Ikegami, A.; Moser, M.; Schiltz, E.; Tokuda, H.; Müller, M. A Derivative of Lipid A Is Involved in Signal-Recognition Particle/SecYEG-Dependent and -Independent Membrane Integrations. *J. Biol. Chem.* **2006**, *281*, 35667–35676.
20. Kawashima, Y.; Miyazaki, E.; Müller, M.; Tokuda, H.; Nishiyama, K. Diacylglycerol Specifically Blocks Spontaneous Integration of Membrane Proteins and Allows Detection of a Factor-Assisted Integration. *J. Biol. Chem.* **2008**, *283*, 24489–24496.

21. Nishiyama, K.; Maeda, M.; Abe, M.; Kanamori, T.; Shimamoto, K.; Kusumoto, S.; Ueda, T.; Tokuda, H. A Novel Complete Reconstitution System for Membrane Integration of the Simplest Membrane Protein. *Biochem. Biophys. Res. Commun.* **2010**, *394*, 733–736.
22. Moser, M.; Nagamori, S.; Huber, M.; Tokuda, H.; Nishiyama, K. Glycolipozyme MPIase Is Essential for Topology Inversion of SecG During Preprotein Translocation. *Proc. Natl. Acad. Sci. U. S. A.* **2013**, *110*, 9734–9739.
23. Nishikawa, H.; Sasaki, M.; Nishiyama, K. *In Vitro* Assay for Bacterial Membrane Protein Integration Into Proteoliposomes. *Bio.-Protoc.* **2020**, *10*, e3626. https://doi.org/10.21769/BioProtoc.3626.
24. Shimizu, Y.; Inoue, A.; Tomari, Y.; Suzuki, T.; Yokogawa, T.; Nishikawa, K.; Ueda, T. Cell-Free Translation Reconstituted With Purified Components. *Nat. Biotechnol.* **2001**, *19*, 751–755.
25. Kiefer, D.; Kuhn, A. Hydrophobic Forces Drive Spontaneous Membrane Insertion of the Bacteriophage Pf3 Coat Protein Without Topological Control. *EMBO J.* **1999**, *18*, 6299–6306.
26. Kuhn, H. M.; Meier-Dieter, U.; Mayer, H. ECA, the Enterobacterial Common Antigen. *FEMS Microbiol. Rev.* **1988**, *4*, 195–222.
27. Fujikawa, K.; Suzuki, S.; Nagase, R.; Ikeda, S.; Mori, S.; Nomura, K.; Nishiyama, K.; Shimamoto, K. Syntheses and Activities of the Functional Structures of a Glycolipid Essential for Membrane Protein Integration. *ACS Chem. Biol.* **2018**, *13*, 2719–2727.
28. Fujikawa, K.; Nishiyama, K.; Shimamoto, K. Enzyme-like Glycolipids MPIase Involved in Membrane Protein Integration of *E. coli*. *Trends Glycosci. Glycotechnol.* **2019**, *31*, E151–E158. https:/doi.org/10.4052/tigg.1705.1E.
29. Paulsen, H.; Lorentzen, J. P. Synthesis of the Immunologically Essential Saccharide Sequence of the Enterobacterial Common Antigen. *Angew. Chem. Int. Ed. Engl.* **1985**, *24*, 773–775.
30. Liu, L.; Zha, J.; DiGiandomenico, A.; McAllister, D.; Stover, C. K.; Wang, Q.; Boons, G.-J. Synthetic Enterobacterial Common Antigen (ECA) for the Development of a Universal Immunotherapy for Drug-Resistant Enterobacteriaceae. *Angew. Chem. Int. Ed.* **2015**, *54*, 10953–10957.
31. Pozsgay, V.; Brisson, J. R.; Jennings, H. J. Synthesis of a Tri- and a Tetra-Saccharide Fragment of the Capsular Polysaccharide of Type III Group B Streptococcus. *Carbohydr. Res.* **1990**, *205*, 133–146.
32. Wang, G.; Lu, Z.; Ding, N.; Zhang, W.; Wang, P.; Li, Y. Differentiating the 2,3-Diols of Glucopyranosides by 4,6-*O*-Benzylidene-Protected 1,2-D-Glucopyranosylorthoesters Strategy. *Carbohydr. Res.* **2011**, *346*, 2368–2373.
33. Nicolaou, K. C.; Mitchell, H. J.; Rodriguez, R. M.; Fylaktkidou, K. C.; Suzuki, H. Total Synthesis of Everninomicin 13,384-1—Part 3: Synthesis of the DE-Fragment and Completion of the Total Synthesis. *Angew. Chem. Int. Ed.* **1999**, *38*, 3345–3350.
34. Bartra, M.; Romea, P.; Urpí, F.; Vilarrasa, J. A Fast Procedure for the Reduction of Azides and Nitro Compounds Based on the Reducing Ability of SN(SR)3 Species. *Tetrahedron* **1990**, *46*, 587–594.
35. Sabesan, S.; Neira, S. Synthesis of Glycosyl Phosphates and Azides. *Carbohydr. Res.* **1992**, *223*, 169–185.
36. Sim, M. M.; Kondo, H.; Wong, C.-H. Synthesis and Use of Glycosyl Phosphites—An Effective Route to Glycosyl Phosphates, Sugar Nucleotides, and Glycosides. *J. Am. Chem. Soc.* **1993**, *115*, 2260–2267.
37. Tsukamoto, H.; Kahne, D. *N*-Methylimidazolium Chloride-Catalyzed Pyrophosphate Formation: Application to the Synthesis of Lipid I and NDP-Sugar Donors. *Bioorg. Med. Chem. Lett.* **2011**, *21*, 5050–5053.

38. Gold, H.; van Delft, P.; Meeuwenoord, N.; Code, J. D. C.; Overkleeft, H. S.; van der Marel, G. A. Synthesis of Sugar Nucleotides by Application Phosphoramidites. *J. Org. Chem.* **2008**, *73*, 9458–9460.
39. Nguyen, H. Q.; Davis, R. A.; Gervay-Hague, J. Synthesis and Structural Characterization of Three Unique *Helicobacter pylori* α-Cholesteryl Phosphatidyl Glucosides. *Angew. Chem. Int. Ed.* **2014**, *53*, 13400–13403.
40. Tanaka, H. Recent Approaches to the Chemical Synthesis of Sugar Nucleoside Diphosphates. *Trends Glycosci. Glycotechnol.* **2015**, *27*, 99–110.
41. Ellis, R. J. Do Molecular Chaperones Have to be Proteins? *Biochem. Biophys. Res. Commun.* **1997**, *238*, 687–692.
42. Mori, S.; Nomura, K.; Fujikawa, K.; Osawa, T.; Shionyu, M.; Yoda, T.; Shirai, T.; Tsuda, S.; Yoshizawa-Kumagaye, K.; Masuda, S.; Nishio, H.; Yoshiya, T.; Suzuki, S.; Muramoto, M.; Nishiyama, K.; Shimamoto, K. Intermolecular Interactions Between a Membrane Protein and a Glycolipid Essential for Membrane Protein Integration. *ACS Chem. Biol.* **2022**, *17*, 609–618.
43. Jason-Moller, L.; Murphy, M.; Bruno, J. Overview of Biocore Systems and Their Applications. *Curr. Protoc. Protein Sci.* **2006**, *45* (1). https://doi.org/10.1002/0471140864.ps1913s45.
44. Markgren, P.-O.; Schaal, W.; Hämäläinen, M.; Karlén, A.; Hallberg, A.; Samuelsson, B.; Danielson, U. H. Relationship Between Structure and Interaction Kinetics for HIV-1 Protease Inhibitors. *J. Med. Chem.* **2002**, *45*, 5430–5439.
45. Mayer, M.; Meyer, B. Characterization of Ligand Binding by Saturation Transfer Difference NMR Spectroscopy. *Angew. Chem. Int. Ed.* **1999**, *38*, 1784–1788.
46. Crooke, E. *Escherichia coli* DnaA Protein-Phospholipid Interactions: In Vivo and In Vitro. *Biochimie* **2001**, *83*, 19–23.
47. Matsumoto, K. Dispensable Nature of Phosphatidylglycerol in *Escherichia coli*: Dual Roles of Anionic Phospholipids. *Mol. Microbiol.* **2001**, *39*, 1427–1433.
48. Breukink, E.; Demel, R. A.; de Korte-Kool, G.; de Kruijff, B. SecA Insertion Into Phospholipids Is Stimulated by Negatively Charged Lipids and Inhibited by ATP—A Monolayer Study. *Biochemistry* **1992**, *31*, 1119–1124.
49. Dowhan, W.; Vitrac, H.; Bogdanov, M. Lipid-Assisted Membrane Protein Folding and Topogenesis. *Protein J.* **2019**, *38*, 274–288.
50. Nomura, K.; Yamaguchi, T.; Mori, S.; Fujikawa, K.; Nishiyama, K.; Shimanouchi, T.; Tanimoto, Y.; Morigaki, K.; Shimamoto, K. Alteration of Membrane Physicochemical Properties by Two Factors for Membrane Protein Integration. *Biophys. J.* **2019**, *117*, 99–110.
51. Fujikawa, K.; Nomura, K.; Nishiyama, K.; Shimamoto, K. Novel Glycolipid Involved in Membrane Protein Integration: Structure and Mode of Action. *J. Synth. Org. Chem. Jpn.* **2019**, 77, 1096–1105.
52. Nomura, K.; Mori, S.; Fujikawa, K.; et al. Role of a Bacterial Glycolipid in Sec-Independent Membrane Protein Insertion. *Sci. Rep.* **2022**, *12*, 12231. https://doi.org/10.1038/s41598-022-16304-1.
53. Kamemoto, Y.; Funaba, N.; Kawakami, M.; Sawasato, K.; Kanno, K.; Suzuki, S.; Nishikawa, H.; Sato, R.; Nishiyama, K. Biosynthesis of Glycolipid MPIase (Membrane Protein Integrase) Is Independent of the Genes for ECA (Enterobacterial Common Antigen). *J. Gen. Appl. Microbiol.* **2020**, *66*, 169–174.
54. Rush, J. S.; Rick, P. D.; Waechter, C. J. Polyisoprenyl Phosphate Specificity of UDP-GlcNAc:Undecaprenyl Phosphate *N*-Acetylglucosaminyl 1-P Transferase from *E. coli*. *Glycobiology* **1997**, 7, 315–322.
55. Rick, P. D.; Hubbard, G. L.; Kitaoka, M.; Nagaki, H.; Kinoshita, T.; Dowd, S.; Simplaceanu, V.; Ho, C. Characterization of the Lipid-Carrier Involved in the Synthesis of Enterobacterial Common Antigen (ECA) and Identification of a Novel

Phosphoglyceride in a Mutant of *Salmonella typhimurium* Defective in ECA Synthesis. *Glycobiology* **1998**, *8*, 557–567.
56. Sawasato, K.; Sato, R.; Nishikawa, H.; Iimura, N.; Kamemoto, Y.; Fujikawa, K.; Yamaguchi, T.; Kuruma, Y.; Tamura, Y.; Endo, T.; Ueda, T.; Shimamoto, K.; Nishiyama, K. CdsA Is Involved in Biosynthesis of Glycolipid MPIase Essential for Membrane Protein Integration In Vivo. *Sci. Rep.* **2019**, *9*, 1372. https://doi.org/10.1038/s41598-018-37809-8.
57. Icho, T.; Sparrow, C. P.; Raetz, C. R. H. Molecular Cloning and Sequencing of the Gene for CDP-Diglyceride Synthetase of *Escherichia coli*. *J. Biol. Chem.* **1985**, *260*, 2078–2083.
58. Sparrow, C. P.; Raetz, C. R. Purification and Properties of the Membrane-Bound CDP-Diglyceride Synthetase From *Escherichia coli*. *J. Biol. Chem.* **1985**, *260*, 2084–2091.
59. Sato, R.; Sawasato, K.; Nishiyama, K. YnbB Is a CdsA Paralogue Dedicated to Biosynthesis of Glycolipid MPIase Involved in Membrane Protein Integration. *Biochem. Biophys. Res. Commun.* **2019**, *510*, 636–642.
60. Tamura, Y.; Harada, Y.; Yamano, K.; Watanabe, K.; Ishikawa, D.; Ohshima, C.; Nishikawa, S.; Yamamoto, H.; Endo, T. Identification of Tam41 Maintaining Integrity of the TIM23 Protein Translocator Complex in Mitochondria. *J. Cell Biol.* **2006**, *174*, 631–637.
61. Sawasato, K.; Suzuki, S.; Nishiyama, K. Increased Expression of the Bacterial Glycolipid MPIase Is Required for Efficient Protein Translocation Across Membranes in Cold Conditions. *J. Biol. Chem.* **2019**, *294*, 8403–8411.
62. Sawasato, K.; Sekiya, Y.; Nishiyama, K. Two-Step Induction of cdsA Promotors Leads to Upregulation of the Glycolipid MPIase at Cold Temperature. *FEBS Lett.* **2019**, *593*, 1711–1721.
63. Pogliano, K. J.; Beckwith, J. The CS Sec Mutants of *Escherichia coli* Reflect the Cold Sensitivity of Protein Export Itself. *Genetics* **1993**, *133*, 763–773.
64. Nishiyama, K.; Mizushima, S.; Tokuda, H. A Novel Membrane-Protein Involved in Protein Translocation Across the Cytoplasmic Membrane of *Escherichia coli*. *EMBO J.* **1993**, *12*, 3409–3415.
65. Scotti, P. A.; Urbanus, M. L.; Brunner, J.; de Gier, J.-W.; Heijne, G.; Does, C.; Driessen, A. J. M.; Oudega, B.; Luirink, J. YidC, the *Escherichia coli* Homologue of Mitochondrial Oxa1p, Is a Component of the Sec Translocase. *EMBO J.* **2000**, *19*, 542–549.
66. Yi, L.; Dalbey, R. E. Oxa1/Alb3/YidC System for Insertion of Membrane Proteins in Mitochondria, Chloroplasts and Bacteria (Review). *Mol. Membr. Biol.* **2005**, *22*, 101–111.
67. Dalbey, R. E.; Kuhn, A.; Zhu, L.; Kiefer, D. The Membrane Insertase YidC. *Biochim. Biophys. Acta* **2014**, *1843*, 1489–1496.
68. Samuelson, J. C.; Chen, M.; Jiang, F.; Möller, I.; Wiedmann, M.; Kuhn, A.; Phillips, G. J.; Dalbey, R. E. YidC Mediates Membrane Protein Insertion in Bacteria. *Nature* **2000**, *406*, 637–641.
69. Nishikawa, H.; Sasaki, M.; Nishiyama, K. Membrane Insertion of F-0 Subunit of F0F1 ATPase Depends on Glycolipozyme MPIase and Is Stimulated by YidC. *Biochem. Biophys. Res. Commun.* **2017**, *487*, 477–482.
70. Sasaki, M.; Nishikawa, H.; Suzuki, S.; Moser, M.; Huber, M.; Sawasato, K.; Matsubayashi, H. T.; Kumazaki, K.; Tsukazaki, T.; Kuruma, Y.; Nureki, O.; Ueda, T.; Nishiyama, K. The Bacterial Protein YidC Accelerates MPIase-Dependent Integration of Membrane Proteins. *J. Biol. Chem.* **2019**, *294*, 18898–18908.
71. Nishikawa, H.; Kanno, K.; Endo, Y.; Nishiyama, K. Ring Assembly of c Subunits of F0F1-ATP Synthase in *Propionigenium modestum* Requires YidC and UncI Following MPIase-Dependent Membrane Insertion. *FEBS Lett.* **2021**, *595*, 647–654.

72. Endo, Y.; Shimizu, Y.; Nishikawa, H.; Sawasato, K.; Nishiyama, K. Interplay Between MPIase, YidC, and PMF During Sec-Independent Insertion of Membrane Proteins. *Life Sci. Alliance* **2022**, *5*, e202101162. https://doi.org/10.26508/lsa.202101162.
73. Kumazaki, K.; Chiba, S.; Takemoto, M.; Furukawa, A.; Nishiyama, K.; Sugano, Y.; Mori, T.; Dohmae, N.; Hirata, K.; Nakada-Nakura, Y.; Maturana, A. D.; Tanaka, Y.; Mori, H.; Sugita, Y.; Arisaka, F.; Ito, K.; Ishitani, R.; Tsukazaki, T.; Nureki, O. Structural Basis of Sec-Independent Membrane Protein Insertion by YidC. *Nature* **2014**, *509*, 516–520.
74. Kumazaki, K.; Kishimoto, T.; Furukawa, A.; Mori, H.; Tanaka, Y.; Dohmae, N.; Ishitani, R.; Tsukazaki, T.; Nureki, O. Crystal Structure of *Escherichia coli* YidC, Membrane Protein Chaperone and Insertase. *Sci. Rep.* **2014**, *4*, 7299. https://doi.org/10.1038/srep07299.
75. Tsukazaki, T. Structural Basis of the Sec Translocon and YidC Revealed Through X-ray Crystallography. *Protein J.* **2019**, *38*, 249–261.
76. Nishiyama, K.; Suzuki, T.; Tokuda, H. Inversion of the Membrane Topology of SecG Coupled With SecA-Dependent Preprotein Translocation. *Cell* **1996**, *85*, 71–81.

Author index

N

O

P

Q

R

S

Subject index

Note: Page numbers followed by "*f*" indicate figures and "*s*" indicate schemes.

N

O

P

R

S

T

U

V

Y

FSC
www.fsc.org
MIX
Papier | Fördert
gute Waldnutzung
FSC® C083411

Kolibri 360 GmbH
Ferdinand-Jühlke-Straße 7
99095 Erfurt, Deutschland
produktsicherheit@kolibri360.de